A Comparison of U.S. Geological Survey Three-Dimensional Model Estimates of Groundwater Source Areas and Velocities to Independently Derived Estimates, Idaho National Laboratory and Vicinity, Idaho

By Jason C. Fisher, Joseph P. Rousseau, Roy C. Bartholomay, and Gordon W. Rattray

DOE/ID-22218
Prepared in cooperation with the U.S. Department of Energy

Scientific Investigations Report 2012–5152

U.S. Department of the Interior
U.S. Geological Survey

U.S. Department of the Interior
KEN SALAZAR, Secretary

U.S. Geological Survey
Marcia K. McNutt, Director

U.S. Geological Survey, Reston, Virginia: 2012

For more information on the USGS—the Federal source for science about the Earth, its natural and living resources, natural hazards, and the environment, visit http://www.usgs.gov or call 1–888–ASK–USGS.

For an overview of USGS information products, including maps, imagery, and publications, visit http://www.usgs.gov/pubprod

To order this and other USGS information products, visit http://store.usgs.gov

Suggested citation:
Fisher, J.C., Rousseau, J.P., Bartholomay, R.C, and Rattray, G.W., 2012, A comparison of U.S. Geological Survey three-dimensional model estimates of groundwater source areas and velocities to independently derived estimates, Idaho National Laboratory and vicinity, Idaho: U.S. Geological Survey Scientific Investigations Report 2012-5152 (DOE/ID-22218), 130 p.

Contents

Figures

Figures—Continued

Tables

Conversion Factors, Datums, and Abbreviations and Acronyms

Conversion Factors

Multiply	By	To obtain
Length		
foot (ft)	0.3048	meter (m)
mile (mi)	1.609	kilometer (km)
Area		
square mile (mi^2)	2.590	square kilometer (km^2)
Flow rate		
foot per day (ft/d)	0.3048	meter per day (m/d)
cubic foot per second (ft^3/s)	0.02832	cubic meter per second (m^3/s)
Hydraulic conductivity		
foot per day (ft/d)	0.3048	meter per day (m/d)

Temperature in degrees Celsius (°C) may be converted to degrees Fahrenheit (°F) as follows:

$$°F=(1.8×°C)+32.$$

Concentrations of chemical constituents in water are given either in milligrams per liter (mg/L) or micrograms per liter (µg/L).

Datums

Vertical coordinate information is referenced to the National Geodetic Vertical Datum of 1929 (NGVD 29).

Horizontal coordinate information is referenced to the North American Datum of 1927 (NAD 27).

The projection Albers used in the North American Datum of 1927, a central meridian of 113°W., standard parallels of 42° 50′ N. and 44° 10′ N., a false easting of 656,166.67 ft, and the latitude of projection's origin was 41° 30′ N.

Altitude, as used in this report, refers to distance above the vertical datum.

Abbreviations and Acronyms

Abbreviation or acronym	Definition
ATRC	Advanced Test Reactor Complex
bls	below land surface
B	boron
BC	Birch Creek
BLR	Big Lost River
CFA	Central Facilities Area
DOE	U.S. Department of Energy
ESRP	eastern Snake River Plain
head	hydraulic head
INL	Idaho National Laboratory
INTEC	Idaho Nuclear Technology and Engineering Center
K	hydraulic conductivity
Li	lithium
LLR	Little Lost River
MFC	Materials and Fuels Complex
ML	Mud Lake
MLMS	Multilevel Monitoring System
Mo	Monteview
NRTS	National Reactor Testing Station
NRF	Naval Reactors Facility
Re	Reno
RMSE	Root-mean-square error
RWMC	Radioactive Waste Management Complex
SDA	Subsurface Disposal Area
TAN	Test Area North
Te	Terreton
USGS	U.S. Geological Survey

A Comparison of U.S. Geological Survey Three-Dimensional Model Estimates of Groundwater Source Areas and Velocities to Independently Derived Estimates, Idaho National Laboratory and Vicinity, Idaho

By Jason C. Fisher, Joseph P. Rousseau, Roy C. Bartholomay, and Gordon W. Rattray

Abstract

The U.S. Geological Survey (USGS), in cooperation with the U.S. Department of Energy, evaluated a three-dimensional model of groundwater flow in the fractured basalts and interbedded sediments of the eastern Snake River Plain aquifer at and near the Idaho National Laboratory to determine if model-derived estimates of groundwater movement are consistent with (1) results from previous studies on water chemistry type, (2) the geochemical mixing at an example well, and (3) independently derived estimates of the average linear groundwater velocity. Simulated steady-state flow fields were analyzed using backward particle-tracking simulations that were based on a modified version of the particle tracking program MODPATH. Model results were compared to the 5-microgram-per-liter lithium contour interpreted to represent the transition from a water type that is primarily composed of tributary valley underflow and streamflow-infiltration recharge to a water type primarily composed of regional aquifer water. This comparison indicates several shortcomings in the way the model represents flow in the aquifer. The eastward movement of tributary valley underflow and streamflow-infiltration recharge is overestimated in the north-central part of the model area and underestimated in the central part of the model area. Model inconsistencies can be attributed to large contrasts in hydraulic conductivity between hydrogeologic zones.

Sources of water at well NPR-W01 were identified using backward particle tracking, and they were compared to the relative percentages of source water chemistry determined using geochemical mass balance and mixing models. The particle tracking results compare reasonably well with the chemistry results for groundwater derived from surface-water sources (–28 percent error), but overpredict the proportion of groundwater derived from regional aquifer water (108 percent error) and underpredict the proportion of groundwater derived from tributary valley underflow from the Little Lost River valley (–74 percent error). These large discrepancies may be attributed to large contrasts in hydraulic conductivity between hydrogeologic zones and (or) a short-circuiting of underflow from the Little Lost River valley to an area of high hydraulic conductivity.

Independently derived estimates of the average groundwater velocity at 12 well locations within the upper 100 feet of the aquifer were compared to model-derived estimates. Agreement between velocity estimates was good at wells with travel paths located in areas of sediment-rich rock (root-mean-square error [RMSE] = 5.2 feet per day [ft/d]) and poor in areas of sediment-poor rock (RMSE = 26.2 ft/d); simulated velocities in sediment-poor rock were 2.5 to 4.5 times larger than independently derived estimates at wells USGS 1 (less than 14 ft/d) and USGS 100 (less than 21 ft/d). The models overprediction of groundwater velocities in sediment-poor rock may be attributed to large contrasts in hydraulic conductivity and a very large, model-wide estimate of vertical anisotropy (14,800).

Introduction

The Idaho National Laboratory (INL) was established by the U.S. Atomic Energy Commission, now the U.S. Department of Energy (DOE), in 1949 to build, operate, and test nuclear reactors. The scope of work at the INL increased from the 1950s through the 1970s to include other nuclear-research programs, the reprocessing of spent nuclear fuel, and the storage and disposal of radioactive waste. More than 50 years of waste disposal associated with nuclear-reactor research and nuclear-fuel reprocessing at the INL has resulted in measurable concentrations of contaminants in the eastern Snake River Plain (ESRP) aquifer beneath the INL.

The INL covers an area of about 890 mi? and overlies the west-central part of the ESRP in southeastern Idaho (fig. 1A). The underlying ESRP aquifer is a major source of water for agricultural, industrial, and domestic use in southeastern Idaho. Wastewater disposal sites at the Test Area North (TAN), Naval Reactors Facility (NRF), Advanced Test

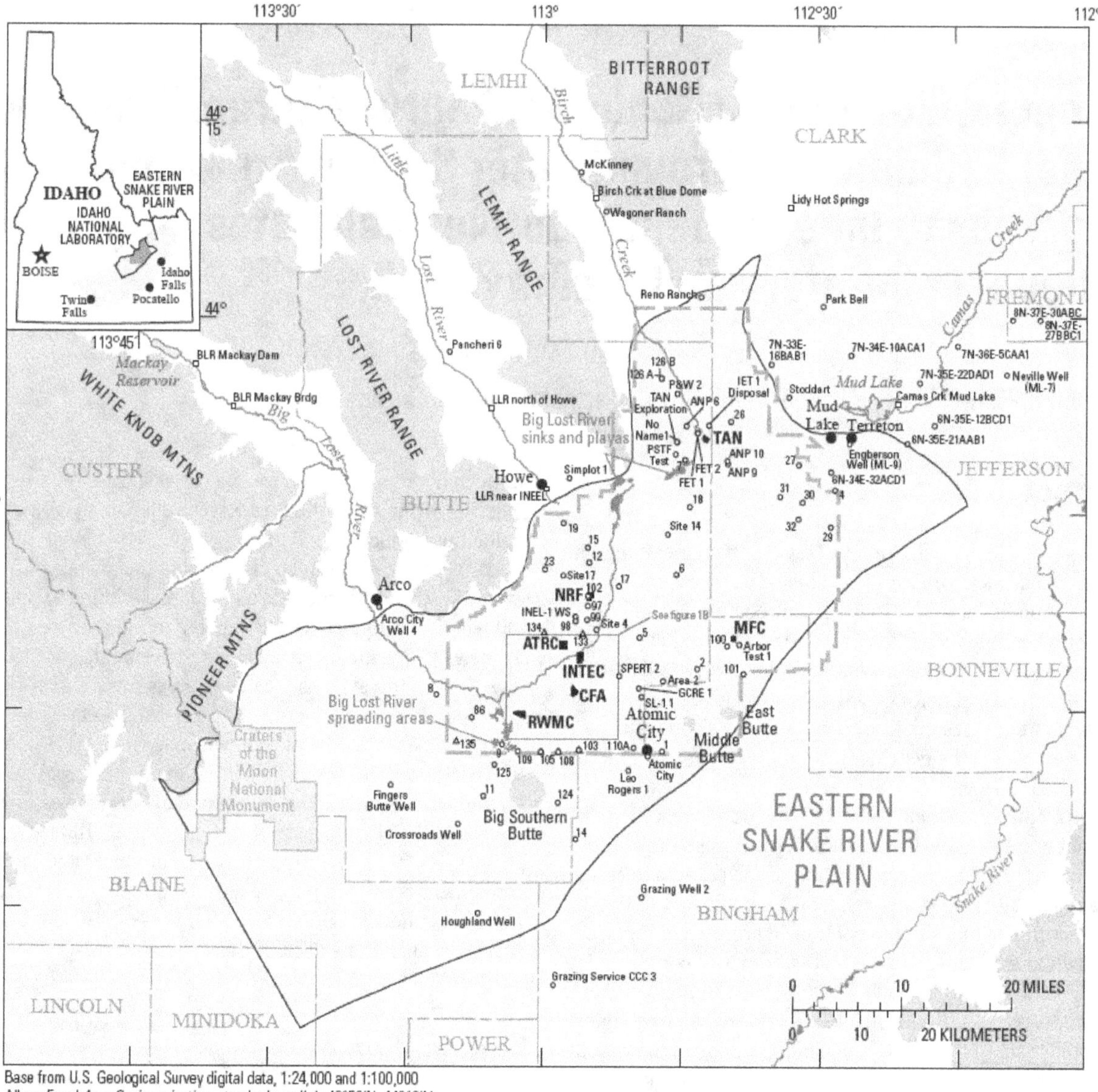

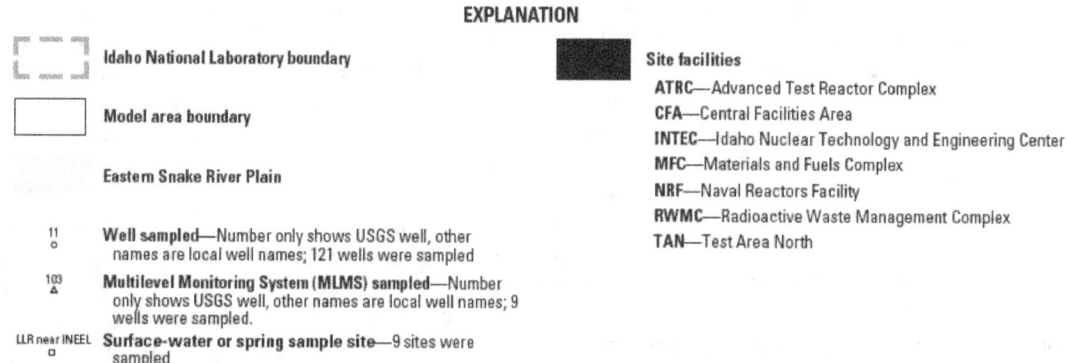

Base from U.S. Geological Survey digital data, 1:24,000 and 1:100,000
Albers Equal-Area Conic projection, standard parallels 42°50'N, 44°10'N;
central meridian 113°00'W; North American Datum of 1927.

A. Idaho National Laboratory and vicinity

EXPLANATION

Idaho National Laboratory boundary

Model area boundary

Eastern Snake River Plain

11 Well sampled—Number only shows USGS well, other
    names are local well names; 121 wells were sampled

103 Multilevel Monitoring System (MLMS) sampled—Number
    only shows USGS well, other names are local well names; 9
    wells were sampled.

LLR near INEEL Surface-water or spring sample site—9 sites were
    sampled

Site facilities
ATRC—Advanced Test Reactor Complex
CFA—Central Facilities Area
INTEC—Idaho Nuclear Technology and Engineering Center
MFC—Materials and Fuels Complex
NRF—Naval Reactors Facility
RWMC—Radioactive Waste Management Complex
TAN—Test Area North

Figure 1. Location of the Idaho National Laboratory, the model area, selected facilities, wells, and key cultural and physiographic features, Idaho National Laboratory and vicinity, Idaho.

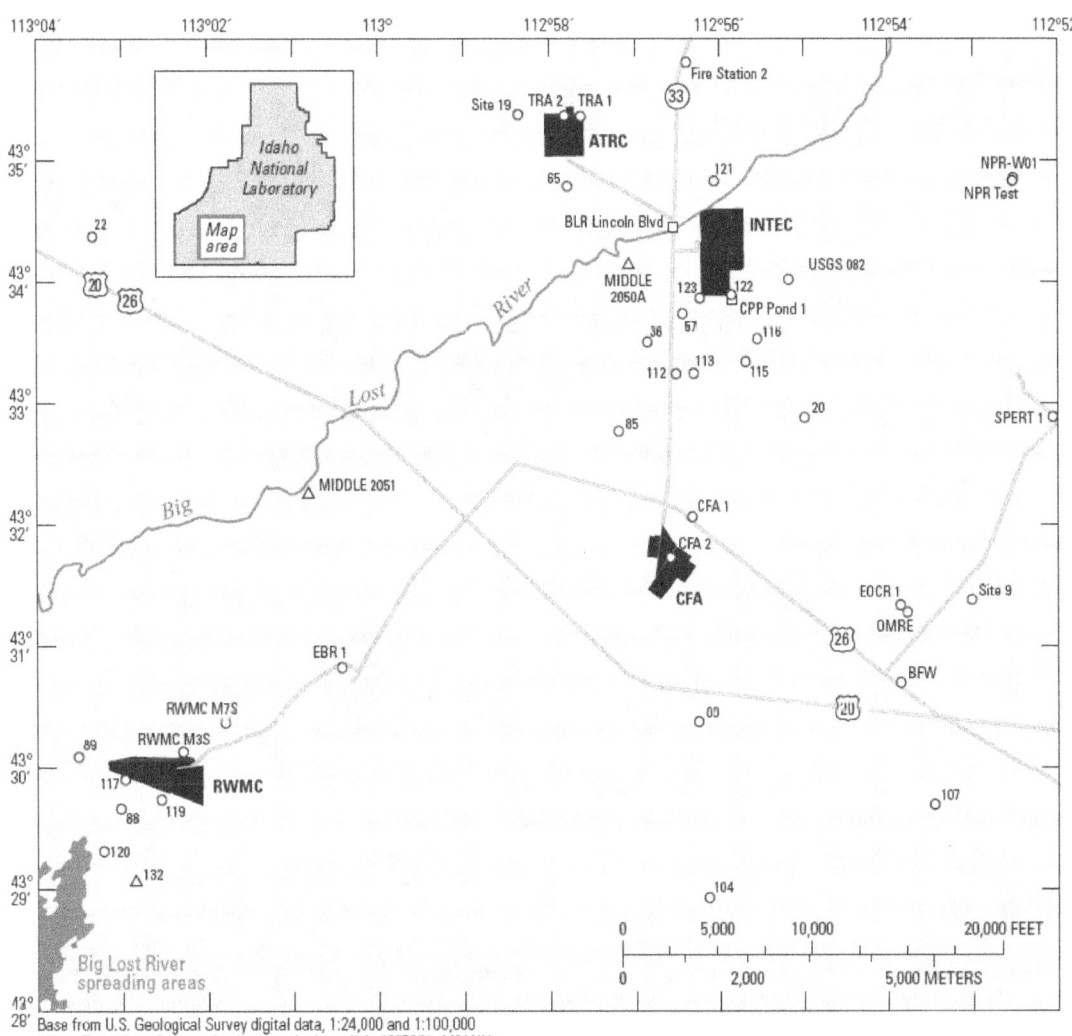

Base from U.S. Geological Survey digital data, 1:24,000 and 1:100,000
Albers Equal-Area Conic projection, standard parallels 42°50′N, 44°10′N;
central meridian 113°00′W; North American Datum of 1927.

B. **Southwest part of Idaho National Laboratory**

EXPLANATION

Site facilities

    ATRC—Advanced Test Reactor Complex

    CFA—Central Facilities Area

    INTEC Idaho Nuclear Technology and Engineering Center

    RWMC Radioactive Waste Management Complex

Eastern Snake River Plain

83
O   **Well sampled**—Number only shows USGS well, other names are local well names

132
△   **Multilevel Monitoring System sampled**—Number only shows USGS well, other names are local well names

CPP Pond 1
☐   **Surface-water or spring sample site**

Figure 1.—Continued

Reactor Complex (ATRC; formerly known as the Reactor Technology Complex, RTC, and the Test Reactor Area, TRA), and Idaho Nuclear Technology and Engineering Center (INTEC; formerly known as the Idaho Chemical Processing Plant, ICPP) (figs. 1*A* and 1*B*) have been primary sources of radioactive and chemical waste contaminants in water from the ESRP aquifer. These wastewater disposal sites have, in the past, included lined evaporation ponds, unlined infiltration ponds and ditches, drain fields, and injection wells. Waste materials buried in shallow pits and trenches within the subsurface disposal area at the Radioactive Waste Management Complex (RWMC) also have been sources of contaminants in groundwater.

Numerical models of steady-state and transient flow were developed by the U.S. Geological Survey (USGS) in cooperation with the DOE to simulate the movement of groundwater in the west-central part of the ESRP aquifer (fig. 1*A*) (Ackerman and others, 2010). These flow models were constructed using the USGS modular, three-dimensional (3-D), finite-difference groundwater flow model, MODFLOW-2000 (Harbaugh and others, 2000), and cover an area of 1,940 mi^2 that includes most of the INL. Steady-state flow was simulated to represent conditions in 1980 using average streamflow infiltration from 1966–80 to represent streamflow-infiltration recharge in the Big Lost River channel, spreading areas, sinks, and playas. The transient flow model simulates groundwater flow between 1980 and 1995, a period that included a 5-year wet cycle (1982–86) followed by an 8-year dry cycle (1987–94).

In these models, the fractured basalts, intercalated beds of fine-grained sediments, and rhyolitic ash flow tuffs of the ESRP aquifer are represented as porous media with nonuniform properties and are grouped into four primary hydrogeologic units. In areas where unconsolidated alluvial sediment constitutes more than 11 percent of the stratigraphic section, these primary hydrogeologic units are further subdivided to distinguish sediment-rich areas (greater than 11 percent sediment) from sediment-poor areas (less than 11 percent sediment). This distinction results in eight hydrogeologic zones to represent the basalts and interbedded sediments and one hydrogeologic zone to represent the rhyolitic rocks (fig. 2).

Three physical and three artificial boundaries define the model area (fig. 3). The physical boundaries are the water table, the northwest mountain front, and the base of the aquifer. The artificial boundaries are the northeast regional underflow, the southeast flowline, and the southwest regional underflow. Inflow to the aquifer is across the water table,

northwest mountain-front, and northeast regional-underflow boundaries. The base of the aquifer and the southeast flowline boundary are treated as no-flow boundaries. Outflow is across the southwest regional-underflow and water table boundaries (table 1).

Groundwater inflow to the aquifer increases progressively in a direction downgradient of the northeast boundary and along the regional direction of groundwater flow from northeast to southwest. This increased flow is the result of tributary valley underflow along the northwest mountain-front boundary and precipitation-, irrigation-, industrial wastewater-, and streamflow-infiltration recharge across the water table boundary. The remainder of the inflow originates from underflow along the northeast boundary. The large inflows from the northwest and northeast boundaries contribute prominently to the chemical character and distribution of the major groundwater types within the model area (figs. 3 and 4).

Depth to the water table ranges from 200 ft in the northern part of the model area to 1,000 ft in the southern part. Depth to the base of the aquifer ranges from 700 to 4,800 ft below land surface (bls) (Ackerman and others, 2010, p. 9). The 3-D geometry of the aquifer is irregular. The interpreted depth to the base of the aquifer indicates large changes in the saturated thickness of the aquifer across the model area. Aquifer thickness generally increases from northeast to southwest and from west to east and is greatest southwest of the INL (fig. 5).

The model grid consists of 0.25 mi × 0.25 mi cells (fig. 3). These cells are projected vertically downwards across six model layers extending from the 1980 water table to the base of the aquifer. From top to bottom these layers are labeled one through six and are of varying thickness: Layer 1 is about 100 ft thick, varying with the water table altitude; Layer 2 is 100 ft thick; Layer 3 is 0 to 100 ft thick; Layer 4 is 0 to 200 ft thick; Layer 5 is 0 to 300 ft thick; and Layer 6 is 0 to 3,229 ft thick (Ackerman and others, 2010, fig. 15). Model layers 3, 4, 5, and 6 are not present everywhere in the model area (fig. 2).

Model-derived estimates of horizontal hydraulic conductivity (K_h; table 2) indicate that K_h varies from less than one to almost 2 orders of magnitude between each sediment-poor (1, 2, 3, 4) and sediment-rich (11, 22, 33, 44) hydrogeologic zone. With the exception of hydrogeologic zone 4, the best-fit estimates of K_h fall within a narrow range of uncertainty, as defined by their upper and lower 95-percent confidence limits. The calibrated value for the horizontal hydraulic conductivity of hydrogeologic zone 4 ($K_{h,4}$) is less certain as indicated by the large difference in the upper and lower 95-percent confidence limits for this parameter.

Table 1. Summary of modeled flows across the northwest mountain-front boundary, northeast regional-underflow boundary, southeast-flowline boundary, water-table boundary, and base of the aquifer boundary used in the U.S. Geological Survey three-dimensional steady-state groundwater flow model, Idaho National Laboratory and vicinity, Idaho.

[Locations of flow boundaries are shown in figure 3. Flows are rounded to nearest tenth. Streamflow reaches are located between the model boundary and streamflow-gaging station 504 (600–601), within the Big Lost River spreading area (602–605), between streamflow-gaging stations 504 and 506 (606–607), within the Big Lost River sinks and playas (608–610), within the Little Lost River downstream from the model boundary (611), and within Birch Creek downstream from the model boundary (612). Identifiers used to locate streamflow-gaging stations and stream reaches are shown in figure 3. **Abbreviation:** ft³/s, cubic foot per second]

Boundaries	Spatial boundary characteristics	Flow (ft³/s)	Percentage of total flow
Inflow boundaries			
Northwest mountain-front boundary	Nonuniform		
Big Lost River valley (BLR)	Uniform	361.0	17.1
Little Lost River valley (LLR)	Uniform	223.0	10.5
Birch Creek valley (BC)	Uniform	62.0	2.9
Mountain-front	Uniform	0.0	0.0
Northwest mountain-front subtotal		**646.0**	**30.5**
Northeast regional-underflow boundary	Nonuniform		
Reno Ranch section (Re)	Nonuniform	76.7	3.6
Monteview section (Mo)	Nonuniform	101.3	4.8
Mud Lake section (ML)	Nonuniform	358.6	16.9
Terreton section (Te)	Nonuniform	688.5	32.5
Northeast regional-underflow subtotal		**1,225.0**	**57.8**
Southeast-flowline boundary	Uniform	0.0	0.0
Water table boundary	Nonuniform		
Precipitation recharge	Uniform	70.0	3.3
Irrigation infiltration	Nonuniform	21.6	1.0
Industrial water use returns	Nonuniform	5.9	0.3
Streamflow infiltration	Nonuniform		
Big Lost River infiltration	Nonuniform		
Stream reach 600–601	Uniform	21.8	1.0
Stream reach 602–605	Uniform	43.0	2.0
Stream reach 606–607	Uniform	24.3	1.2
Stream reach 608–610	Uniform	36.2	1.7
Little Lost River infiltration			
Stream reach 611	Uniform	3.0	0.2
Birch Creek infiltration			
Stream reach 612	Uniform	20.0	1.0
Streamflow-infiltration subtotal		**148.3**	**7.1**
Base of the aquifer boundary	Uniform	0.0	0.0
Total inflow		**2,116.8**	**100.0**
Outflow boundaries			
Southwest regional-underflow boundary	Nonuniform	2,072.0	97.9
Water table boundary	Nonuniform		
Irrigation well discharge	Nonuniform	37.2	1.7
Industrial well discharge	Nonuniform	7.6	0.4
Total outflow		**2,116.8**	**100.0**

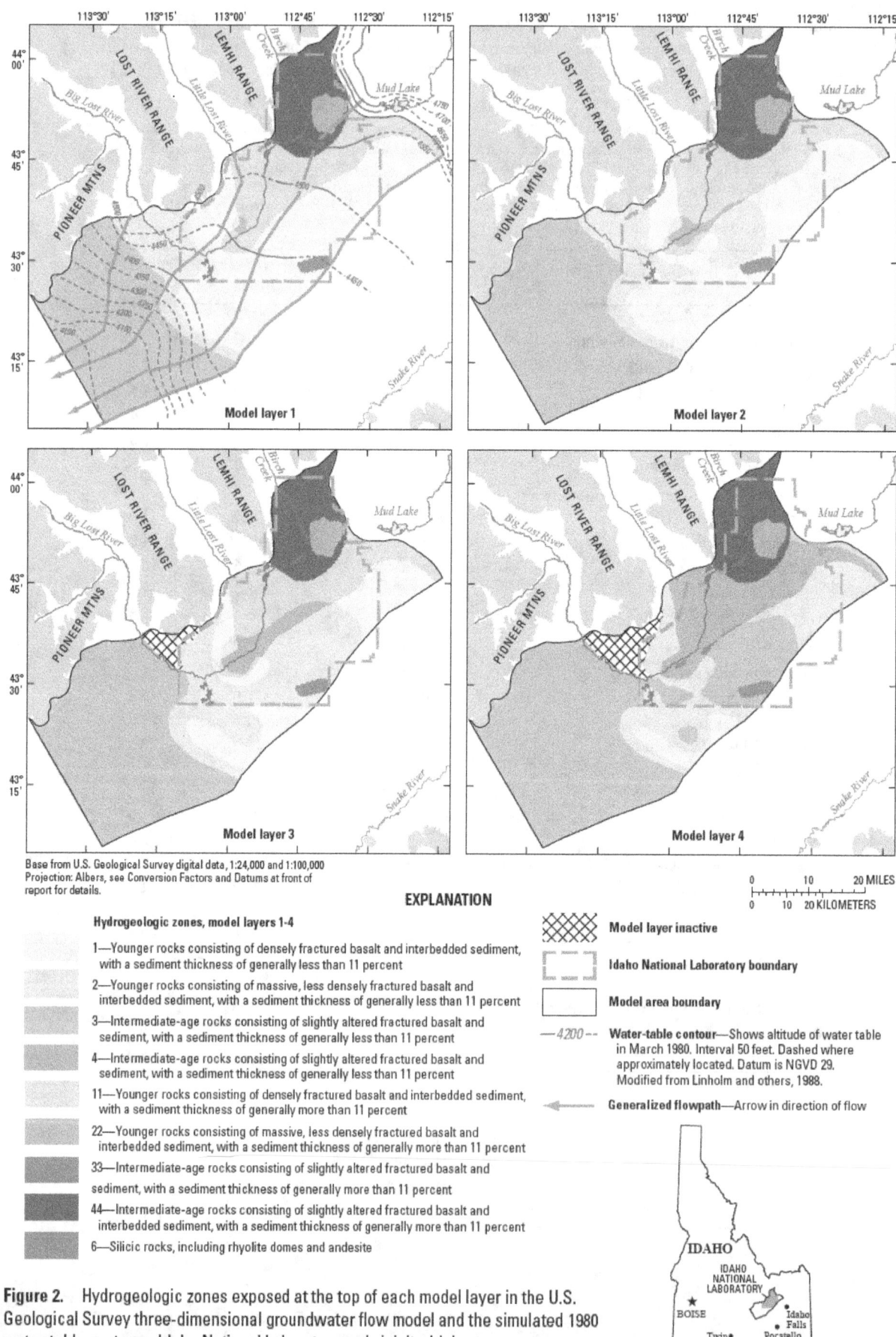

Base from U.S. Geological Survey digital data, 1:24,000 and 1:100,000
Projection: Albers, see Conversion Factors and Datums at front of
report for details.

EXPLANATION

Hydrogeologic zones, model layers 1-4

1—Younger rocks consisting of densely fractured basalt and interbedded sediment, with a sediment thickness of generally less than 11 percent

2—Younger rocks consisting of massive, less densely fractured basalt and interbedded sediment, with a sediment thickness of generally less than 11 percent

3—Intermediate-age rocks consisting of slightly altered fractured basalt and sediment, with a sediment thickness of generally less than 11 percent

4—Intermediate-age rocks consisting of slightly altered fractured basalt and sediment, with a sediment thickness of generally less than 11 percent

11—Younger rocks consisting of densely fractured basalt and interbedded sediment, with a sediment thickness of generally more than 11 percent

22—Younger rocks consisting of massive, less densely fractured basalt and interbedded sediment, with a sediment thickness of generally more than 11 percent

33—Intermediate-age rocks consisting of slightly altered fractured basalt and sediment, with a sediment thickness of generally more than 11 percent

44—Intermediate-age rocks consisting of slightly altered fractured basalt and interbedded sediment, with a sediment thickness of generally more than 11 percent

6—Silicic rocks, including rhyolite domes and andesite

Model layer inactive

Idaho National Laboratory boundary

Model area boundary

— 4200 — — Water-table contour—Shows altitude of water table in March 1980. Interval 50 feet. Dashed where approximately located. Datum is NGVD 29. Modified from Linholm and others, 1988.

Generalized flowpath—Arrow in direction of flow

Figure 2. Hydrogeologic zones exposed at the top of each model layer in the U.S. Geological Survey three-dimensional groundwater flow model and the simulated 1980 water-table contours, Idaho National Laboratory and vicinity, Idaho.

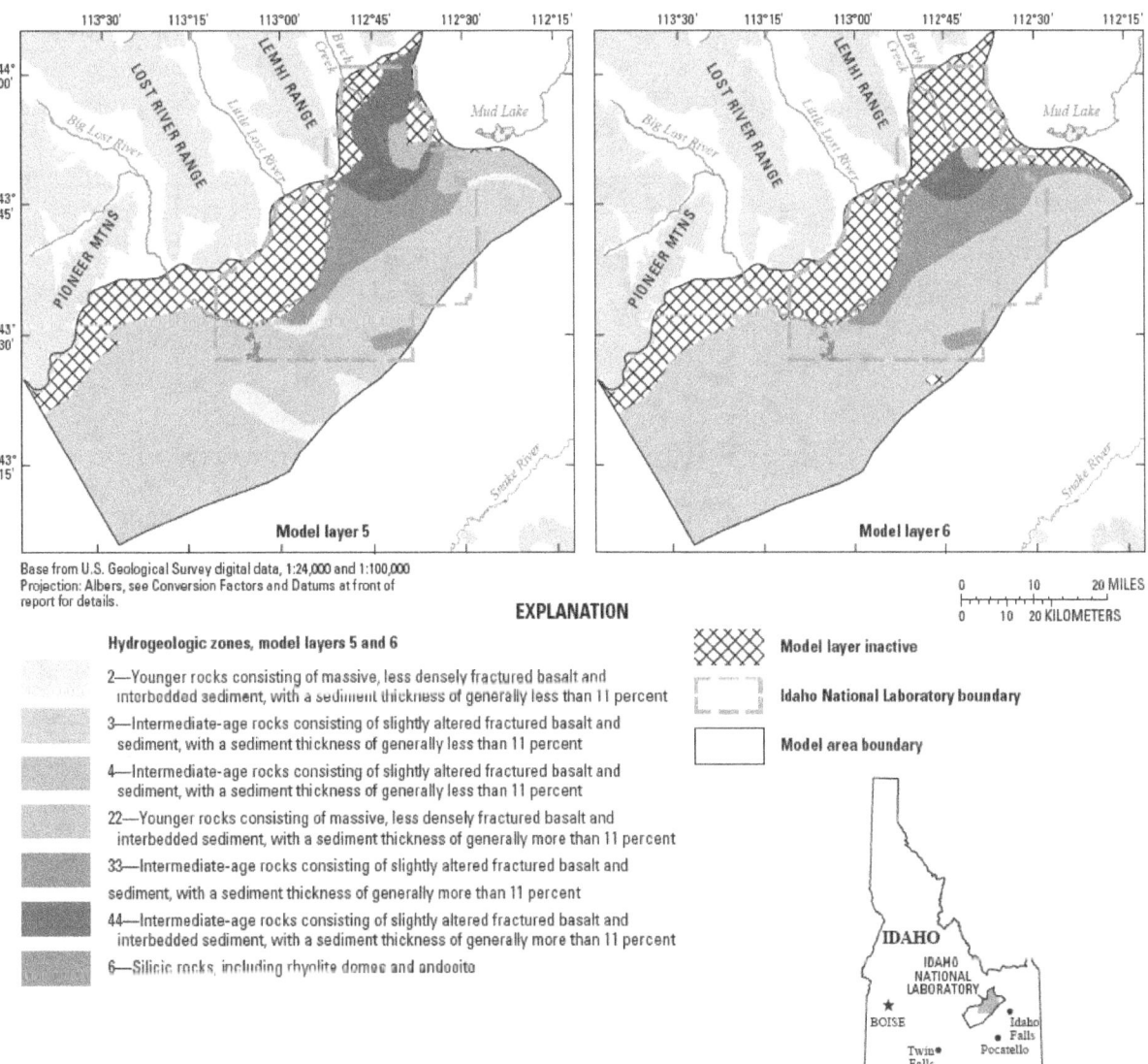

Base from U.S. Geological Survey digital data, 1:24,000 and 1:100,000
Projection: Albers, see Conversion Factors and Datums at front of
report for details.

EXPLANATION

Hydrogeologic zones, model layers 5 and 6

2—Younger rocks consisting of massive, less densely fractured basalt and interbedded sediment, with a sediment thickness of generally less than 11 percent

3—Intermediate-age rocks consisting of slightly altered fractured basalt and sediment, with a sediment thickness of generally less than 11 percent

4—Intermediate-age rocks consisting of slightly altered fractured basalt and sediment, with a sediment thickness of generally less than 11 percent

22—Younger rocks consisting of massive, less densely fractured basalt and interbedded sediment, with a sediment thickness of generally more than 11 percent

33—Intermediate-age rocks consisting of slightly altered fractured basalt and sediment, with a sediment thickness of generally more than 11 percent

44—Intermediate-age rocks consisting of slightly altered fractured basalt and interbedded sediment, with a sediment thickness of generally more than 11 percent

6—Silicic rocks, including rhyolite domes and andesite

Model layer inactive

Idaho National Laboratory boundary

Model area boundary

Figure 2.—Continued

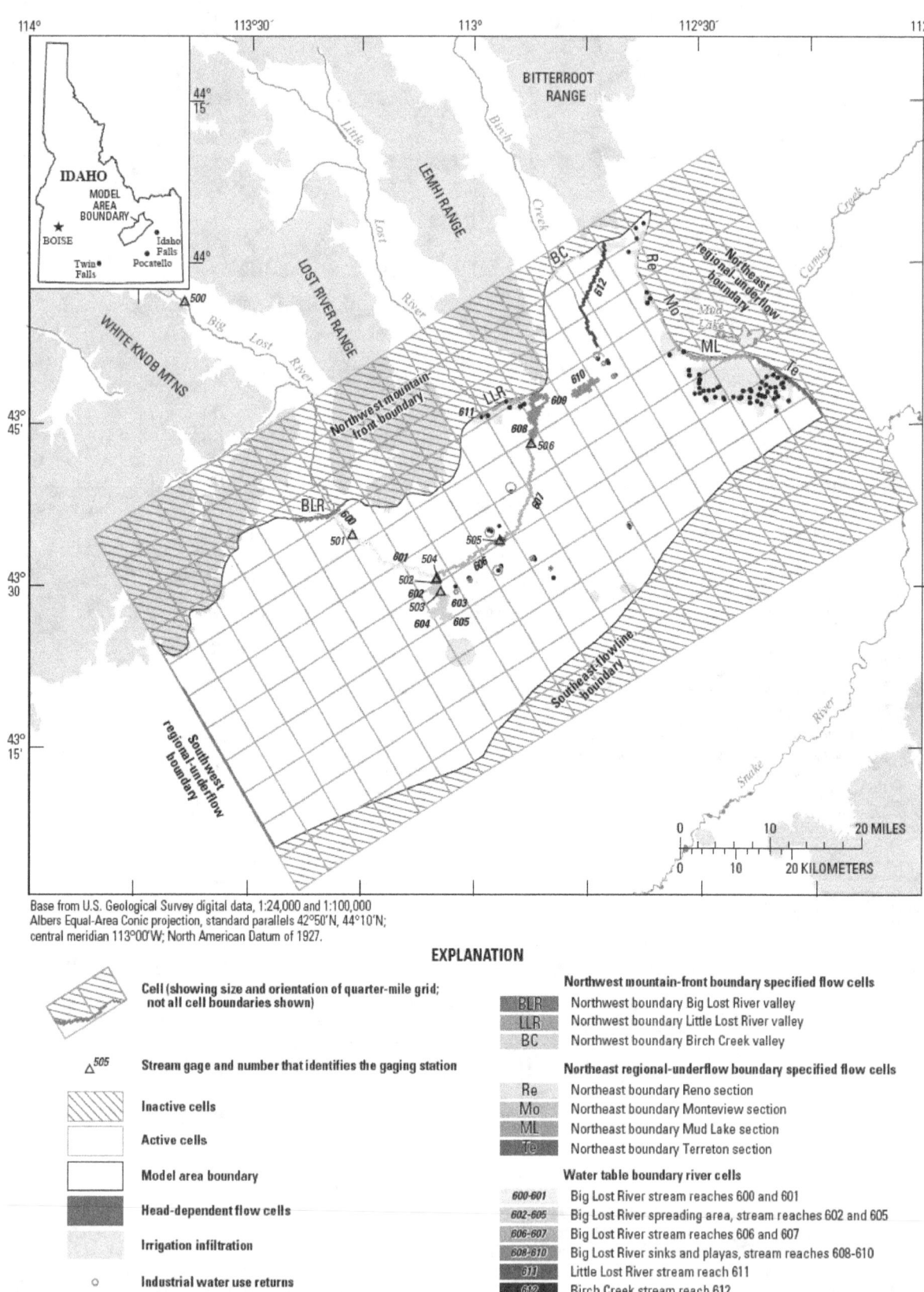

Base from U.S. Geological Survey digital data, 1:24,000 and 1:100,000
Albers Equal-Area Conic projection, standard parallels 42°50'N, 44°10'N;
central meridian 113°00'W; North American Datum of 1927.

EXPLANATION

Cell (showing size and orientation of quarter-mile grid;
not all cell boundaries shown)

△505   Stream gage and number that identifies the gaging station

Inactive cells

Active cells

Model area boundary

Head-dependent flow cells

Irrigation infiltration

○   Industrial water use returns

•   Irrigation and industrial well discharge

Northwest mountain-front boundary specified flow cells

BLR   Northwest boundary Big Lost River valley
LLR   Northwest boundary Little Lost River valley
BC   Northwest boundary Birch Creek valley

Northeast regional-underflow boundary specified flow cells

Re   Northeast boundary Reno section
Mo   Northeast boundary Monteview section
ML   Northeast boundary Mud Lake section
Te   Northeast boundary Terreton section

Water table boundary river cells

600-601   Big Lost River stream reaches 600 and 601
602-605   Big Lost River spreading area, stream reaches 602 and 605
606-607   Big Lost River stream reaches 606 and 607
608-610   Big Lost River sinks and playas, stream reaches 608-610
611   Little Lost River stream reach 611
612   Birch Creek stream reach 612

Figure 3.   Spatial discretization of the U.S. Geological Survey three-dimensional groundwater flow model and location of groundwater inflow and outflow boundaries, Idaho National Laboratory and vicinity, Idaho.

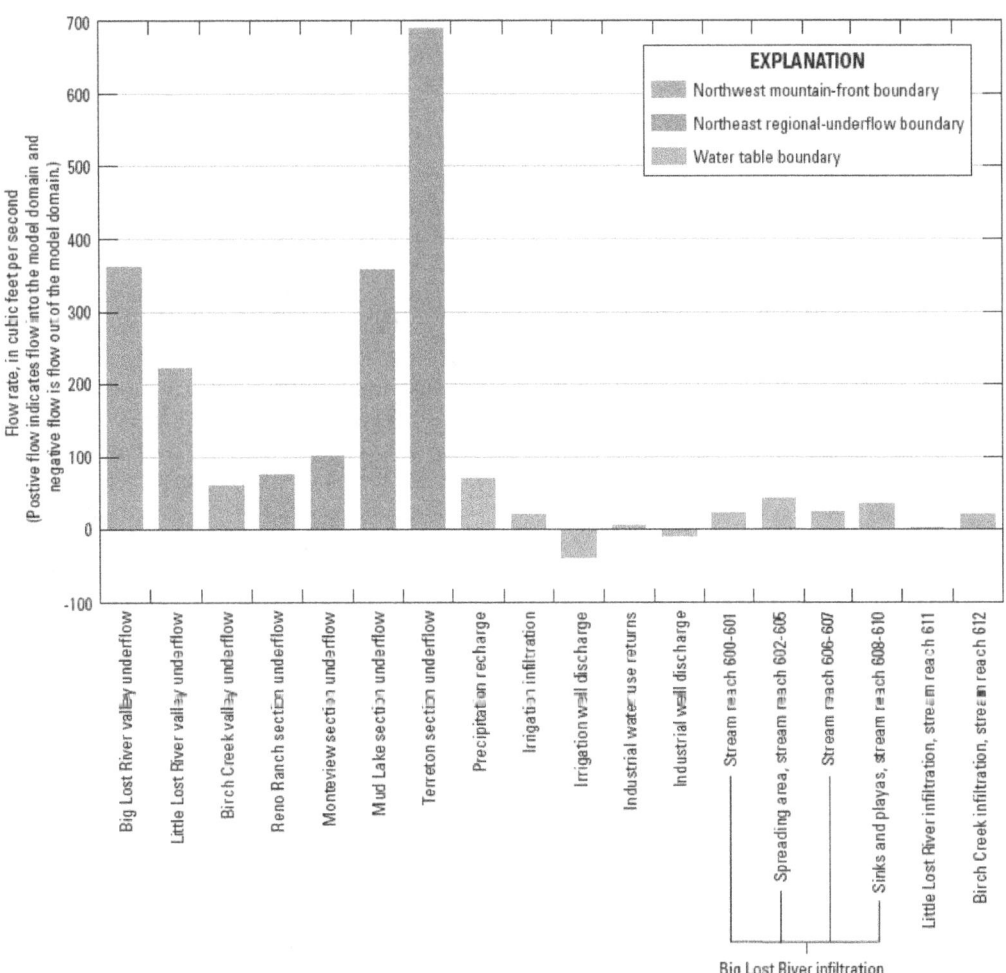

Figure 4.   Inflows and outflows of the steady-state model boundary components. Outflow across the southwest regional-underflow boundary is not shown.

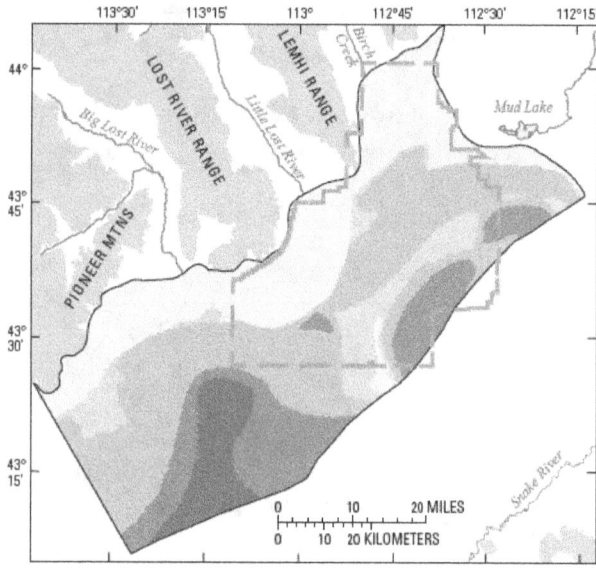

Base from U.S. Geological Survey digital data, 1:24,000 and 1:100,000
Albers Equal-Area Conic projection, standard parallels 42°50'N, 44°10'N;
central meridian 113°00'W; North American Datum of 1927.

EXPLANATION

Aquifer thickness—in feet

200 to 700	**Model area boundary**
701 to 1,200	**Idaho National Laboratory boundary**
1,201 to 2,000	
2,001 to 3,000	
3,001 to 4,031	

Figure 5.   Aquifer thickness as defined by borehole data and surface-based electrical-resistivity surveys, Idaho National Laboratory and vicinity, Idaho.

The K_h for hydrogeologic zone 22, a sediment-rich zone, is larger than its expected range and is an order of magnitude larger than that of hydrogeologic zone 2, a sediment-poor zone. This result is inconsistent with the original conceptual interpretation of how sediment should affect this parameter and could not be readily explained (Ackerman and others, 2010, p. 58). $K_{h,6}$ and $K_{h,33}$ could not be estimated using optimization techniques and were established on the basis of initial trial-and-error modeling (Ackerman and others, 2010, p. 55, and table 8). The uncertainties associated with $K_{h,6}$ and $K_{h,33}$ are unknown. Parameter values for $K_{h,6}$ and $K_{h,33}$ are specified (fixed) in the steady-state model simulations that were used to produce best-fit estimates for the other eight steady-state parameters. The large contrasts in hydraulic conductivity between hydrogeologic zones indicate that simulated flow in the aquifer can be expected to be characterized by (1) sharp refraction across boundaries between sediment-rich and sediment-poor hydrogeologic zones, and (2) horizontal groundwater velocities that differ by at least an order of magnitude for flow in a sediment-rich zone compared to flow in its sediment-poor counterpart.

Vertical anisotropy (table 2), the ratio of horizontal to vertical hydraulic conductivity, is represented by a single model-wide value for all hydrogeologic zones. The use of a single value for vertical anisotropy implies no difference in the anisotropy characteristics (ratio) of sediment-rich and sediment-poor zones. The best-fit estimate of 14,800 for this parameter exceeds its maximum expected value of 1,700 by nearly an order of magnitude. The upper and lower 95-percent confidence limits for this parameter also are large and the lower confidence limit of 7,550 is larger than its maximum expected value, suggesting that there is considerable uncertainty associated with this best-fit estimate. The very large vertical anisotropy indicates that simulated flow in the aquifer can be expected to be characterized by (1) flow that is predominantly horizontal within each hydrogeologic zone, and (2) horizontal velocity components that are several thousand to as much as fifteen thousand times greater than the vertical velocity component within each hydrogeologic zone.

Table 2. Estimates of hydraulic properties, expected intervals, and 95-percent confidence intervals for each of the hydrogeologic zones used in the U.S. Geological Survey three-dimensional steady-state groundwater flow model, Idaho National Laboratory and vicinity, Idaho.

[Hydraulic conductivity in feet per day. Vertical anisotropy is dimensionless and defined as the ratio of horizontal to vertical hydraulic conductivity. Specific yield is dimensionless and defined for each hydrogeologic zone at the water table in model layer 1. Effective porosity is dimensionless. **Zone identifier:** used to locate hydrogeologic zones on maps located in figure 2. **Abbreviations:** –, not available; NA, not applicable]

Hydraulic property	Zone identifier	Expected interval		Estimated value	95-percent confidence interval	
		Lower limit	Upper limit		Lower limit	Upper limit
Hydraulic conductivity	1	0.01	24,000	[1]11,700	10,200	13,500
	2	6.5	1,400	[1]384	244	610
	3	0.32	24,000	[1]435	377	500
	4	0.32	24,000	[1]9,890	1,730	54,700
	6	–	–	[3]86	–	–
	11	0.01	24,000	[1]227	179	296
	22	6.50	1,400	[1]4,780	3,610	6,140
	33	–	–	[3]86	–	–
	44	0.32	24,000	[1]285	225	365
Vertical anisotropy	1, 2, 3, 4, 6, 11, 22, 33, 44	30	1,700	[1]14,800	7,550	29,100
Specific yield	1	0.01	0.30	[2]0.072	0.068	0.077
	2	0.01	0.30	[2]0.115	0.099	0.133
	3	0.01	0.30	[2]0.055	0.039	0.078
	4, 6	0.01	0.30	[6]0.05	–	–
	11	0.01	0.30	[2]0.072	0.066	0.077
	22	0.01	0.30	[6]0.15	–	–
	33	0.01	0.30	[5]NA	–	–
	44	0.01	0.30	[2]0.028	0.023	0.035
Effective porosity	1	–	–	[4]0.07	–	–
	2	–	–	[4]0.14	–	–
	3	–	–	[4]0.03	–	–
	4	–	–	[4]0.05	–	–
	6	–	–	[4]0.05	–	–
	11	–	–	[4]0.07	–	–
	22	–	–	[4]0.15	–	–
	33	–	–	[4]0.05	–	–
	44	–	–	[4]0.03	–	–

[1]Value determined from steady-state model calibration (Ackerman and others, 2010, table 9).

[2]Value determined from transient model calibration (Ackerman and others, 2010, table 15).

[3]Value determined from initial trial-and-error modeling (Ackerman and others, 2010, table 9).

[4]Value determined from large-scale model-derived values of specific yield (Ackerman and others, 2010, table 15), small-scale measurements of bulk or total porosities on individual core samples (Ackerman and others, 2010, table 3), and literature derived estimates of porosity for similar rock types (Freeze and Cherry, 1979, p. 162).

[5]No parameter because hydrogeoloic zone 33 is absent in model layer 1.

[6]Value reflects an assumption that the specific yield of basalts in hydrogeoloic zones 4 and 6 is small and that the specific yield of hydrogeologic zone 22 is large because of the presence of abundant sediment in this zone (Ackerman and others, 2010, p. 77).

Estimated values of effective porosity (table 2) are based on best-fit estimates of specific yield from calibration of the transient model (hydrogeologic zones 1, 2, 3, 11, and 44), small-scale measurements of bulk and total porosities on individual core samples, and literature-derived estimates of porosity for similar rock types (hydrogeologic zones 4, 6, and 44). Effective porosity is the percentage of interconnected pore space in a rock volume and normally is less than total porosity and greater than or equal to the specific yield. Uncertainties in the estimates of effective porosity will have only a minor effect on simulated groundwater velocities compared to uncertainties associated with the model-derived estimates of hydraulic conductivity and vertical anisotropy.

Purpose, Scope, and Methods

This report presents the results of a study to compare model-derived estimates with independently derived estimates of groundwater source areas and groundwater velocities in the west-central part of the ESRP aquifer. This study was done to determine if the representation of flow in the aquifer simulated by the USGS 3-D model is consistent with independent lines of evidence. The USGS 3-D model is based on many assumptions, approximations, and simplifications to model flow in a fractured basalt aquifer that is characterized by extreme heterogeneity and anisotropy. Calibration of this model produced best-fit estimates of hydraulic properties (hydraulic conductivity, vertical anisotropy, and specific yield) that minimized differences between simulated and observed hydraulic heads (head). Although the calibration process produces mathematically precise results, the reliability of those results depends on the validity of user-defined input parameters (for example, boundary fluxes, structure of the geologic framework, steady-state assumption, and heads) and user-specified calibration constraints (for example, a single universal vertical anisotropy). Furthermore, the calibration process does not guarantee a unique solution (that is, different combinations of parameter values could match the observations equally well). Therefore, the relevance of model results cannot be established independently of an assessment of how the integrated effects of all input parameters compare to field observations that are independent of the input parameters used to construct the model.

Backward particle tracking, using the calibrated USGS 3-D steady-state flow model (Ackerman and others, 2010) and a modified version of the particle-tracking program MODPATH (Pollock, 1994; appendix A), were used to (1) trace the sources of groundwater in the model area back to the point where groundwater enters the flow system (crosses a model inflow boundary) to compare with source area water chemistry and the likely geochemical evolution of groundwater along flow paths determined from geochemical studies, and (2) estimate groundwater velocities within the model area. Transient flow simulations were excluded from this analysis because of the added level of complexity associated with reconciling the transient nature of independently derived estimates of groundwater source area and velocity. Source-area water chemistry, concentrations of the trace elements lithium (Li), boron (B), and fluoride (F$^-$) and concentrations of the dissolved gas helium (He) were used to evaluate the backward particle-tracking simulations. The chemistry of Li and B was used to define a mixing transition zone between water types and was compared with model simulation results. Tritium/helium-3 (^{3}H/^{3}He) based estimates of the age of the young fraction of groundwater were used to evaluate model-derived estimates of groundwater velocities.

In this study, three particle-release scenarios were used to simulate groundwater source areas and groundwater velocities:

1. Aerially uniform releases of a single particle in model cells centered at 0.25 mi spacings within each of the six model layers. This backward particle-tracking simulation was used to identify the contributing source areas of groundwater and to estimate groundwater velocities within each model layer at each particle release location.

2. Internally distributed multiple particle releases in model layers 1 and 2 in an area centered at the location of well NPR-W01 in the south-central part of the INL. This backward particle-tracking simulation was used to evaluate mixing of groundwater within the upper 200 ft of the aquifer near a boundary that is interpreted to mark the separation of groundwater derived primarily from tributary valley underflow and streamflow-infiltration recharge, from groundwater derived primarily from regional aquifer underflow.

3. Internally distributed multiple particle releases in model layer 1 at the location of 23 monitoring wells that penetrate less than 100 ft of the aquifer. This backward particle-tracking simulation was used to compare model-derived estimates of average groundwater velocities to independently derived estimates of average linear velocities of the young fraction of groundwater at the particle release location.

In the steady-state model, model layer 1 cells were used to simulate inflow across the water table boundary from streamflow infiltration, precipitation, irrigation return flow, and wastewater disposal (fig. 3). Inflows across the water table boundary represent as much as 11.7 percent of the modeled inflow in the steady-state model and are primarily concentrated in the Big Lost River channel, sinks, playas, and spreading areas (table 1), locations considered to be important sources of rapid, focused recharge and young water in the west-central part of the INL.

An initial analysis of backward particle tracking indicated that particle pathlines rarely terminated at water-table boundary river cells. River cells are assigned a directional component of flow across the top cell face using IFACE, a parameter used by MODPATH, and specified in the MODFLOW stress package files. The insensitivity of particle tracking simulations to IFACE is attributed to (1) an artificially small vertical velocity component across the top cell face resulting from a course representation of the river boundary where streamflow-infiltration recharge is distributed over a 0.25- by 0.25-mi cell face; and (2) a low vertical hydraulic conductivity, the product of a very high single model-wide value of vertical anisotropy, that inhibits upward movement (backward tracking) of particles and prevents them from terminating at the top cell face. Particles entering a 100-ft-thick river cell typically pass through to adjacent cells, and only those particles entering near the top of the river cell terminate at the upper boundary. To better represent streamflow-infiltration recharge in the backward particle-tracking simulation, the MODPATH package (version 5) was modified to account for this process (appendix A). In the modified version of MODPATH, the stopping criterion for particles was changed to include a weak-source evaluation for all particles entering a cell. Weak-source cells describe the case where some of the water flowing out of a cell originates from an internal source (or recharge across an external boundary of the model) and some is passing through the cell from adjacent cells. Because there is no way to know whether backwardly tracked particles entering a weak-source cell should stop at the internal source or pass through the cell to an adjacent cell, an approximation of particle behavior in the cell is necessary. For the particle-tracking simulations presented here, particles were terminated upon entering cells in which recharge was larger than half of the total outflow from the cell. Implementation of this stopping criteria in MODPATH was achieved by setting FRAC, a user-defined input variable, equal to 0.5. A sensitivity analysis of particle-tracking results to variations in FRAC is included in this report.

Additional information on individual wells and streamflow-gaging stations used in this study are found in appendix B and C. Maps showing the locations of these sites are in figures 1A and 1B.

Groundwater Chemistry and Geographic Source Areas of Groundwater

As modeled, 11.7 percent of the groundwater inflow to the model area originates as surface-water infiltration across the water table boundary from precipitation (3.3 percent), streamflow infiltration (7.1 percent), irrigation return flow (1.0 percent), and industrial wastewater return flow (0.3 percent); 30.5 percent as alluvial aquifer underflow along the northwest boundary from the tributary valleys of the Big Lost River (17.1 percent), Little Lost River (10.5 percent), and Birch Creek (2.9 percent); and 57.8 percent as regional aquifer underflow along the northeast boundary (table 1). Each of these water sources has distinctive chemical and geochemical characteristics (Olmsted, 1962; Robertson and others, 1974; Busenberg and others, 2001) that can be used to identify the primary source of water at particle release locations within the upper 200 ft of the aquifer in the model area.

Type A, Type B, and Type C Waters

Olmsted (1962) defined and mapped the distribution of three[1] principal water types in the model area based on the cationic and anionic compositions of water from 96 wells that penetrate the upper 200 ft of the aquifer and on spatial trends in these data. Olmsted's mapping of water types was restricted to water within about the upper 200 ft of the aquifer. Supporting data included chemical analyses of water from wells deeper than 200 ft. However, because of the limited number of wells deeper than 200 ft, the results of these deeper analyses were not included in Olmsted's mapped distributions of water types. In some instances the chemistry of deeper water in a well was compositionally different from shallower water in the same well, indicating limited vertical mixing in these wells and that waters derived from different sources can retain the chemical identity of their geographic source area and coexist in the same well. In other instances, the chemistry of water deeper than 200 ft was the same as that of water in nearby shallower wells, indicating that water from the same source can persist to considerable depth in the aquifer or that mixing occurs through a large part of the upper part of the aquifer. Mapped water types included naturally occurring water and water contaminated from irrigation return flow and industrial wastewater disposal.

---

[1] A fourth water type, Type D with greater than 30 percent sulfate, also was identified by Olmstead and represents contaminated perched water from five wells at the ATRC and a "peculiar" occurrence of shallow aquifer water at ANP 3 (TAN) with a very high sulfate concentration and the lowest concentration of dissolved constituents of all the waters sampled (Olmsted, 1962, p. 24). Olmsted did not map this water type.

In areas unaffected by irrigation return flow, wastewater disposal, or water with high chloride concentrations from "… as yet not identified" natural sources, bicarbonate waters within the upper 200 ft of the aquifer were shown, in most cases, to possess a uniform chemistry with lower specific conductance than water much deeper in the aquifer (Olmsted, 1962, p. 16, 26–31). Within the upper 200 ft of the aquifer, total dissolved solids range from 143 to 273 parts per million (ppm) and average slightly more than 200 ppm; pH is mildly alkaline and ranges from 7.2 to 9.5 and is close to 8.0 in most samples (Olmsted, 1962, p. 17). Olmsted concluded that "The fairly consistent chemical character of most waters is believed to indicate moderately uniform chemical and physical characteristics of the water-bearing rocks throughout the region that includes the NRTS" (Idaho National Laboratory, formerly the National Reactor Testing Station [NRTS]).

Olmsted's classification of water types in the model area is based on the reactive percentages, in milliequivalents per liter, of the cations: calcium plus magnesium (Ca + Mg) and sodium plus potassium (Na + K); the reactive percentages of the anions: bicarbonate and carbonate (HCO_3 + CO_3, note: carbonate is absent in most samples), sulfate plus chloride plus nitrate plus fluoride (SO_4 + Cl + NO_3 + F); and the absolute concentration of silica (SiO_2) in solution (Olmsted, 1962, figs. 7, 8, 9, and 10). Spatial patterns and trends in the proportions of these ionized and un-ionized (SiO_2) constituents, in many cases reflecting only subtle areal variations in relative percentages and concentrations, were used to map the distribution of three water types (appendix D and fig. 6). Olmsted referred to these water types as Types A, B, and C using the classification criteria:

1. Type A: 70 to 100 percent HCO_3 + CO_3; 85 to 100 percent Ca + Mg; and 15 percent or less Na + K,

2. Type B: 70 to 100 percent HCO_3 + CO_3; less than 85 percent Ca + Mg; and more than 15 percent Na + K, and

3. Type C: less than 70 percent HCO_3 + CO_3; with no limits on the proportions of other cations; 30 percent or less SO_4, and generally greater than 20 percent Cl + NO_3 + F.

Type A and Type B waters contain primarily calcium, magnesium, and bicarbonate, and the primary difference between the two is the reactive percentage of Na + K relative to that of Ca + Mg. The average cation percentage of Na + K in Type A water is about 10 percent and in Type B water the average is more than 20 percent (Olmsted, 1962, p. 22). Type A and Type B waters are the dominant water types and are present in the upper 200 ft of the aquifer beneath most of the INL. In some places these waters underlie a thin lens of fresher groundwater that may be from a few feet to as much at 50 ft thick. Olmsted attributed these thin lenses to local recharge from precipitation and runoff into closed drainage basins, and to streamflow infiltration, noting that the very shallow-water chemistry may not represent the primary geographic source area of the groundwater underlying these

thin lenses of fresher water. Olmsted (1962) also indicated that stratification of water types was demonstrated in several wells where different types of sampling methods were used (thief, bailer, or pump).

For samples collected with a bailer, water fills into the bailer as it is lowered through the water column and samples represent a mixture of water in the column to the depth lowered. Thief samples are collected by lowering the sampler to a certain depth and then opening the sampler and collecting water from that depth. Pumped samples pull water from the entire length of an open interval in a well and represent a mixture of any water types that may be present in the water column. Stratification of water types has more recently been demonstrated by sample collection from multilevel monitoring systems (MLMS) installed at the INL (Bartholomay and Twining, 2010). Some stratification of water types may occur throughout the INL, but collection of samples from pumped sites limits the ability to determine the extent of stratification.

As defined by Olmsted:

1. Type A water originates as alluvial-aquifer underflow from the Paleozoic limestones, dolomites, and minor shales forming the Pioneer, Lost River, Lemhi, and Bitterroot mountain ranges and the tributary valleys of the Big Lost River, Little Lost River, and Birch Creek;

2. Type B water originates from the basaltic lava flows and the andesitic and rhyolitic volcanic rocks north and east of the model area. Silicic volcanic rocks in the source areas north of the model area have a much higher content of sodium, potassium, and silica than the carbonate rocks in the source areas of Type A water (Olmsted, 1962, p. 38); and

3. Type C water, less extensive than Types A or B, occurs locally in areas contaminated by wastewater disposal and irrigation return flow and as isolated occurrences of uncertain origin that are surrounded by either Type A or Type B water.

Olmsted's mapped distributions of Type A and Type B water included several anomalous occurrences of Type B water in wells (USGS 7 and USGS 17) within the region mapped as Type A water (fig. 6) and several occurrences where water of more than one type was recovered from the same well (appendix D). Olmsted attributed these anomalous occurrences to (1) vertical variations in water type (USGS 7, USGS 15, and USGS 17) and (2) temporal variations resulting from varying amounts of aquifer inflow from different source areas, particularly downgradient of waste-disposal facilities (USGS 20 and CFA 2, fig. 1B), downgradient of agricultural areas (USGS 30), and along the boundary separating Type A and Type B waters (USGS 5 and USGS 6). In the latter case Olmsted noted that the percentage of Na + K in wells along the boundary separating Type A and Type B water increases when most of the recharge is from the northeast and decreases when most is from the northwest (Olmsted, 1962, p. 26).

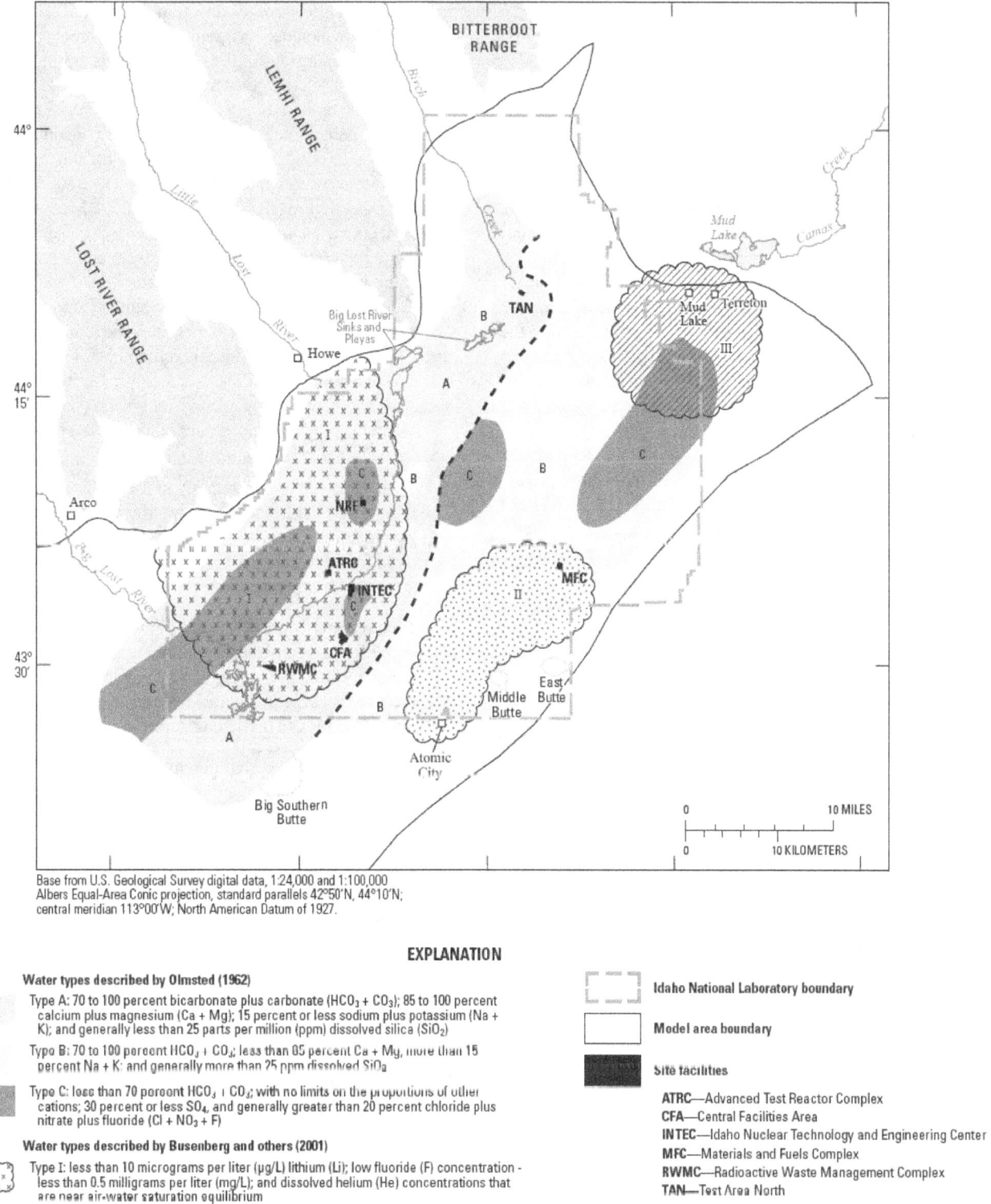

Base from U.S. Geological Survey digital data, 1:24,000 and 1:100,000
Albers Equal-Area Conic projection, standard parallels 42°50'N, 44°10'N;
central meridian 113°00'W; North American Datum of 1927.

EXPLANATION

Water types described by Olmsted (1962)

A — Type A: 70 to 100 percent bicarbonate plus carbonate ($HCO_3 + CO_3$); 85 to 100 percent calcium plus magnesium (Ca + Mg); 15 percent or less sodium plus potassium (Na + K); and generally less than 25 parts per million (ppm) dissolved silica (SiO_2)

B — Type B: 70 to 100 percent $HCO_3 + CO_3$; less than 85 percent Ca + Mg, more than 15 percent Na + K; and generally more than 25 ppm dissolved SiO_2

C — Type C: less than 70 percent $HCO_3 + CO_3$; with no limits on the proportions of other cations; 30 percent or less SO_4, and generally greater than 20 percent chloride plus nitrate plus fluoride ($Cl + NO_3 + F$)

Water types described by Busenberg and others (2001)

I — Type I: less than 10 micrograms per liter (µg/L) lithium (Li); low fluoride (F) concentration - less than 0.5 milligrams per liter (mg/L); and dissolved helium (He) concentrations that are near air-water saturation equilibrium

II — Type II: more than 10 µg/L Li; high F concentrations - from about 0.2 to about 0.8 mg/L; and moderate dissolved He concentrations that are up to about three times air-water saturation equilibrium

III — Type III: more than 10 µg/L Li; high F concentrations - from about 0.2 to about 0.8 mg/L; and high dissolved He conentrations that are greater than about three times air-water saturation equilibrium

- - - - Idaho National Laboratory boundary

Model area boundary

Site facilities

ATRC—Advanced Test Reactor Complex
CFA—Central Facilities Area
INTEC—Idaho Nuclear Technology and Engineering Center
MFC—Materials and Fuels Complex
RWMC—Radioactive Waste Management Complex
TAN—Test Area North

- - - - Major interface between types A and B water

Figure 6. Distribution of Type A, Type B, and Type C waters, outliers of Type B water within Type A water, and the generalized distribution of Type I, Type II, and Type III waters, Idaho National Laboratory and vicinity, Idaho.

Olmsted also noted several locations near TAN (FET 2 and ANP 10), east of the Central Facilities Area (CFA) (EOCR 1, GCRE 1, and OMRE), and at the ATRC (MTR 2) where water in wells deeper than 200 ft possessed a water chemistry that was similar to the chemistry of water in nearby shallower wells (appendix D). Many of these deeper wells are in close proximity to the line separating Type A from Type B water (figs. 1A, 1B, and 6).

Although Olmsted mapped a boundary separating Type A from Type B water (fig. 6) he found "...the gradual rather than abrupt southeastward increase in the percentage of sodium plus potassium and the concentration of silica ... difficult to explain. Mixing of waters from different sources by diffusion and dispersion in basalt aquifers could account for the gradual change; but the transition belt resulting from these processes might be expected to be much narrower than that observed" (Olmsted, 1962, p. 40). Olmsted did not provide an explanation for this conjecture, but did recommend it as an important goal for future hydrochemical study.

Olmsted did not draw attention to the only deep well (SPERT 2, 1,217 ft deep) with Type B water in close proximity to a shallow well with Type A water (SPERT 1, 653 ft deep) (figs. 1A and 1B and appendix D). This single documented occurrence of two different water types in close proximity to each other suggests, at least locally, that mixing of Type A and Type B water may actually occur within a narrow "transition belt" as originally hypothesized by Olmsted (1962).

Type I, Type II, and Type III Waters

Using a different set of classification criteria, Busenberg and others (2001, p. 68, figs. 33 and 34) identified three naturally occurring water types based on the concentrations of the trace elements Li and F, and the dissolved gas He. Busenberg referred to these water types as Types I, II, and III using the classification criteria:

1. Type I: less than 10 micrograms per liter (μg/L) Li; low F concentration—less than about 0.5 milligrams per liter (mg/L); and dissolved He concentrations that are near air-water saturation equilibrium;

2. Type II: more than 10 μg/L Li; high F concentrations—from about 0.2 to about 0.8 mg/L; and moderate dissolved He concentrations that are up to about three times air-water saturation equilibrium; and

3. Type III: more than 10 μg/L Li; high F concentrations—from about 0.2 to about 0.8 mg/L; and high dissolved He concentrations that are greater than three times air-water saturation equilibrium. An exception is water from well USGS 4, which contained background He concentrations.

Although Busenberg's classification criteria are based primarily on the dissolved gas He and on Li and F, chemical signatures of naturally occurring water that he viewed as less likely to be affected by the additions of wastewater or agricultural return flows (Busenberg and others, 2001, p. 9),

many other environmental tracers and naturally occurring trace elements also were used to complement his classification scheme. These included terrigenic He as a percentage of total He; concentrations of the trace elements boron (B) and strontium (Sr); oxygen-18 ($\delta^{18}O$) and carbon-13 ($\delta^{13}C$) isotope ratios; carbon-14 (^{14}C) activity; concentrations of the environmental tracers tritium (^3H), sulfur hexafluoride (SF_6), and chlorofluorocarbons (CFC-11, -12, and -113); and recharge temperatures calculated from concentrations of the dissolved gases nitrogen (N_2) and argon (Ar). Many of these other tracers were used to identify recharge mechanisms leading to mixtures of old and young water and to estimate the age of the young fraction of groundwater.

Busenberg did not map distinct boundaries to separate one water type from another but chose to generalize the distribution of water types based on their chemical, isotopic, and trace element similarities, and their proximity and logical association to contributing groundwater source areas.

As defined by Busenberg and others (2001, p. 68, and fig. 33):

1. Type I water is present mainly in the western part of the INL, is derived from alluvial-aquifer underflow along the northwest boundary and includes contributions from rapid focused recharge from the Big Lost River, spreading areas, sinks, and playas;

2. Type II water is present mainly in the southeastern part of the INL and represents a binary mixture of very old regional water and young water from locally derived rapid focused recharge in the area between the Materials and Fuels Complex (MFC) and Atomic City; and

3. Type III water is present mainly in the northeastern part of the INL and is derived from regional aquifer underflow that is mixed with rapid focused recharge, slow diffuse areal recharge through the unsaturated zone, and agricultural return flow from the Mud Lake and Terreton areas.

Within the upper 200 ft of the aquifer, the distribution of Type I water generally coincides with the distribution of Type A water; and the distributions of Type II and Type III waters generally coincide with the distribution of Type B water (fig. 6).

Mixing Zone Between Tributary Valley Water and Regional Aquifer Water

The trace elements Li and B in conjunction with the water-typing classifications of Olmsted and Busenberg can be used to describe a mixing zone separating groundwater within the upper 200 ft of the aquifer that is derived primarily from tributary valley underflow and streamflow-infiltration recharge (Type A water of Olmsted and Type I water of Busenberg) from groundwater that is derived primarily from regional aquifer underflow (Type B water of Olmsted and Types II and III waters of Busenberg). The approach used to describe this mixing zone assumes that there are no physical or chemical processes (for example, rock-water interactions involving ion exchange, absorption, chemical precipitation, or dissolution) that selectively add or remove Li or B to or from groundwater within the boundaries of the model area.

Concentrations of Lithium and Boron in Tributary Valley Water and Regional Aquifer Water

Lithium and boron concentrations for both surface water and groundwater in the study area are shown in figures 7A and 7B using data from groundwater and surface-water reconnaissance studies of Spinazola and others (1992), Liszewski and Mann (1993), Knobel and others (1999), Busenberg and others (2000), Bartholomay and others (2001), Swanson and others (2003), and recently acquired data from nine wells configured with multilevel monitoring systems (MLMS) (Bartholomay and Twining, 2010) (appendix E).

Concentrations of Li and B are higher in regional groundwater northeast and east of the INL and lower in the alluvial aquifers and streams west of the INL and in the streams and lakes north and east of the INL (Busenberg and others, 2001, p. 9, figs. 3 and 4; figs. 7A and 7B). Locally, very high concentrations of both trace elements are associated with water from Lidy Hot Springs north of the INL. Slightly higher B concentrations are discernible in tributary valley groundwater associated with irrigation return flow in the lower reaches of the Little Lost River valley and downgradient of where this water enters the ESRP aquifer (Busenberg and others, 2001, p. 9, and fig. 4).

Figure 8 presents the distribution over the aquifer of the estimated values of the Li concentration, according to a multilevel B-splines approximation (Lee and others, 1997). The interpolated surface is based on Li concentrations from water samples collected from all wells located in the model domain and open to the upper 200 ft of the aquifer (the number of samples [n] =72). Contour lines of Li concentration greater than about 5 µg/L generally parallel the northeast to southwest regional flow direction. The distribution of Li concentrations is dominated by low concentrations, less than 5 µg/L, along the northwest mountain-front boundary, with a sharp increase in concentrations to the southeast. Interpolated Li concentrations range from 0.8 to 35.4 µg/L.

Figures 9 and 10 present the distribution over the aquifer of the estimated values of the B concentration and the B to Li (B/Li) concentration ratio, according to a multilevel B-splines approximation (Lee and others, 1997), respectively. The interpolated surfaces are based on B concentrations and B/Li concentration ratios from water samples collected from all wells located in the model domain and open to the upper 200 ft of the aquifer. Error in the surface spline interpolation of B concentrations and B/Li concentration ratios is relatively large, in comparison to the interpolation error of Li, due to the small precision (integer precision) and population size (n=58) of B measurements. Interpolated B concentrations range from 12 to 52 µg/L with B concentrations generally

higher east of the line separating Type A and Type B waters, greater than about 30 µg/L. The B concentrations have a markedly heterogeneous distribution with no clear pattern emerging to indicate water type separation (fig. 9).

The distribution of B/Li concentration ratios is dominated by small concentration ratios, less than 4, along the southeast-flowline boundary, with a gradual increase in concentration ratios to the northwest. Contour lines of B/Li concentration ratio less than about 4 generally parallel the northeast to southwest regional flow direction, whereas contour lines of B/Li concentration ratios greater than 4 are heterogeneously distributed. Interpolated B/Li concentration ratios range from 1 to 12 (fig. 10).

A plot of Li concentrations for groundwater and thermal spring water in the study area (n=87), ordered from the lowest to highest concentration, indicates a sharp break in the rank-ordered distribution of Li beginning at a concentration of about 5 µg/L (fig. 11). This plot includes Li measurements from all wells in the study area with a corresponding B measurement, regardless of depth. To avoid spatial biasing, this plot does not include Li measurements from the 41 isolated intervals in the 9 MLMS-instrumented wells and Li measurements at surface-water and wastewater disposal sites shown in figure 7 (appendix E). Water with Li concentrations less than about 5 µg/L occurs west of the line shown in figure 6, and water with Li concentrations greater than about 5 µg/L occurs east of the line. Exceptions to this generalization are the (1) elevated concentrations of Li at well USGS 7, a 1,200-ft-deep well classified as Type B water with a Li concentration of 27 µg/L; (2) slightly elevated concentrations of Li at well USGS 126B, a 472-ft-deep well with a Li concentration of 5.5 µg/L immediately adjacent to well USGS 126A, a 648-ft-deep well with a Li concentration of 4.7 µg/L; (3) slightly elevated Li concentrations at wells USGS 85, USGS 122, and USGS 123, wells near the INTEC waste-disposal site with Li concentrations of 6 µg/L; and (4) USGS 117, a well near the buried waste at the RWMC with a Li concentration of 6.1 µg/L. Surface-water Li concentrations are generally low, less than 5 µg/L, and probably are a major contributor of Li to the aquifer west of the 5 µg/L line and a source of Li dilution in the aquifer east of the 5 µg/L line. The 5 µg/L Li line extends from the northern to the southern end of the model area (fig. 8), is parallel to the regional direction of groundwater flow (fig. 2), and closely approximates the position of the line that Olmsted used to separate Type A and Type B waters (figs. 6 and 8). The westward divergence of the 5 µg/L Li contour in the vicinity of INTEC indicates possible anthropogenic sources of Li at the waste-disposal sites. Although the INL has not shown Li disposal in its wastewater records (Bartholomay and others, 1997, p. 44), Li and its compounds have important application in nuclear energy and high energy fuels (Warde, 1972, p. 662) so slightly elevated Li concentrations could be a byproduct of site operations at the INL. The elevated Li concentration of 11 µg/L in 1991 samples from the CPP Pond 1 (Liszewski and Mann, 1993) and production well CPP 1 (Knobel and others, 1999) at INTEC support this concept.

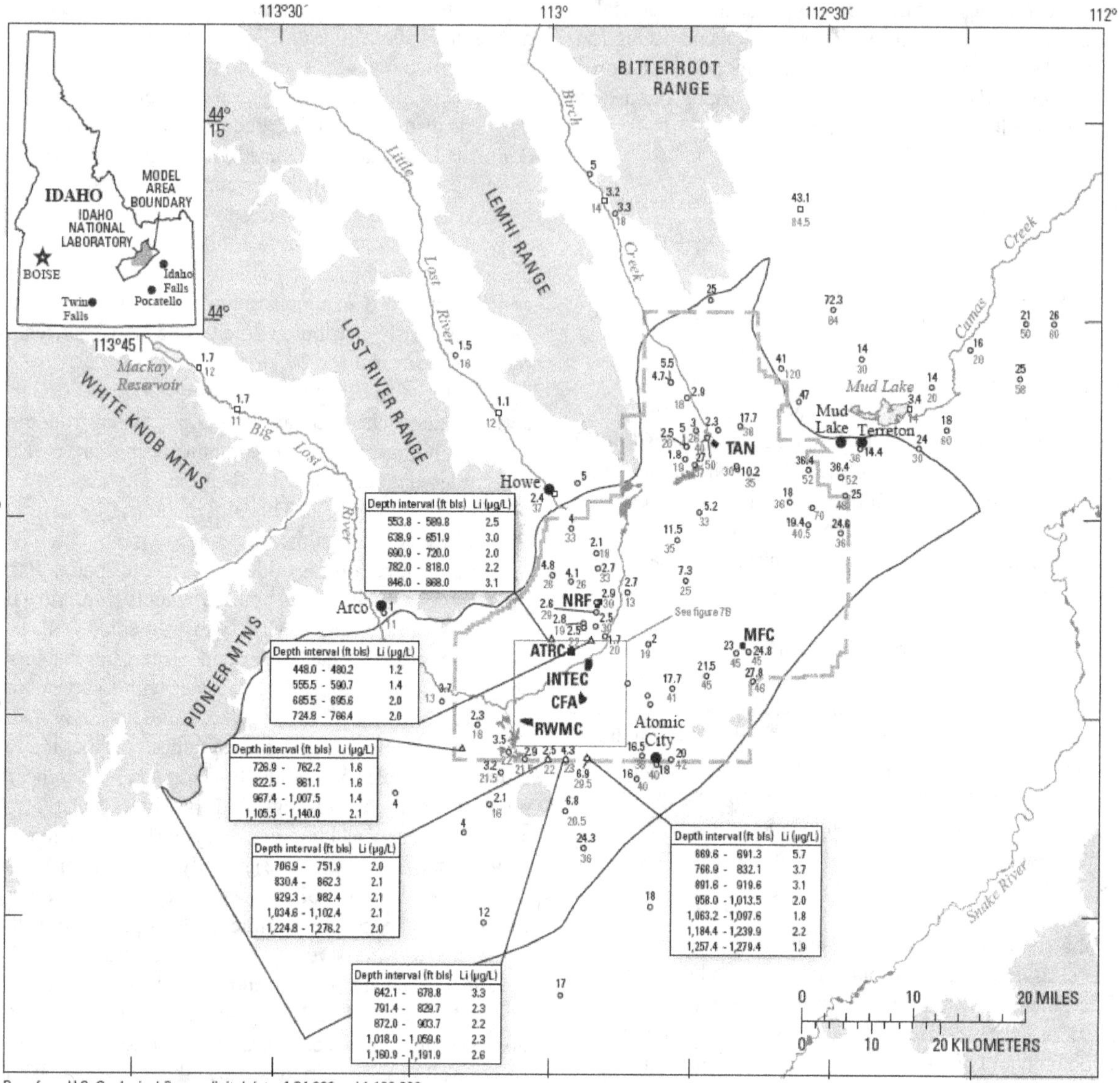

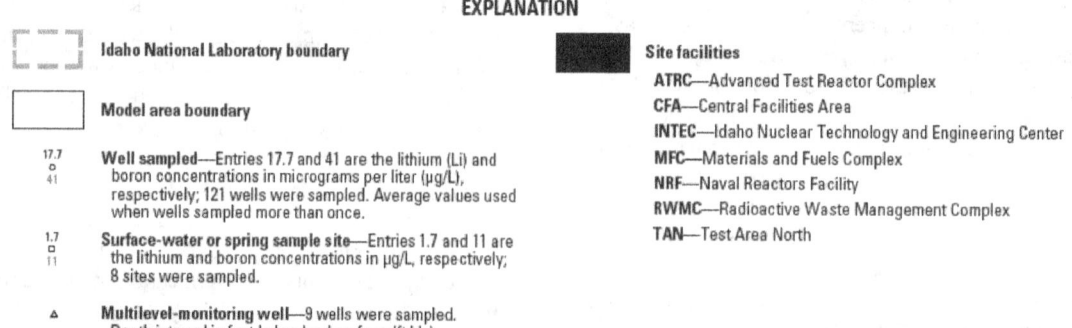

Base from U.S. Geological Survey digital data, 1:24,000 and 1:100,000
Albers Equal-Area Conic projection, standard parallels 42°50'N, 44°10'N;
central meridian 113°00'W; North American Datum of 1927.

A. **Idaho National Laboratory and vicinity**

EXPLANATION

- ⌐ ¬ **Idaho National Laboratory boundary**

- ▢ **Model area boundary**

- 17.7
  ∘   **Well sampled**—Entries 17.7 and 41 are the lithium (Li) and
  41   boron concentrations in micrograms per liter (µg/L),
           respectively; 121 wells were sampled. Average values used
           when wells sampled more than once.

- 1.7
  ∘   **Surface-water or spring sample site**—Entries 1.7 and 11 are
  11   the lithium and boron concentrations in µg/L, respectively;
           8 sites were sampled.

- △   **Multilevel-monitoring well**—9 wells were sampled.
         Depth interval in feet below land surface (ft bls).

■ **Site facilities**
   ATRC—Advanced Test Reactor Complex
   CFA—Central Facilities Area
   INTEC—Idaho Nuclear Technology and Engineering Center
   MFC—Materials and Fuels Complex
   NRF—Naval Reactors Facility
   RWMC—Radioactive Waste Management Complex
   TAN—Test Area North

Figure 7.   Concentrations of lithium and boron in groundwater from 114 wells, 9 wells instrumented with multilevel monitoring systems, 7 surface-water sites, 1 thermal spring, and 1 waste-disposal pond, Idaho National Laboratory and vicinity, Idaho. Concentrations of boron are not available for every site.

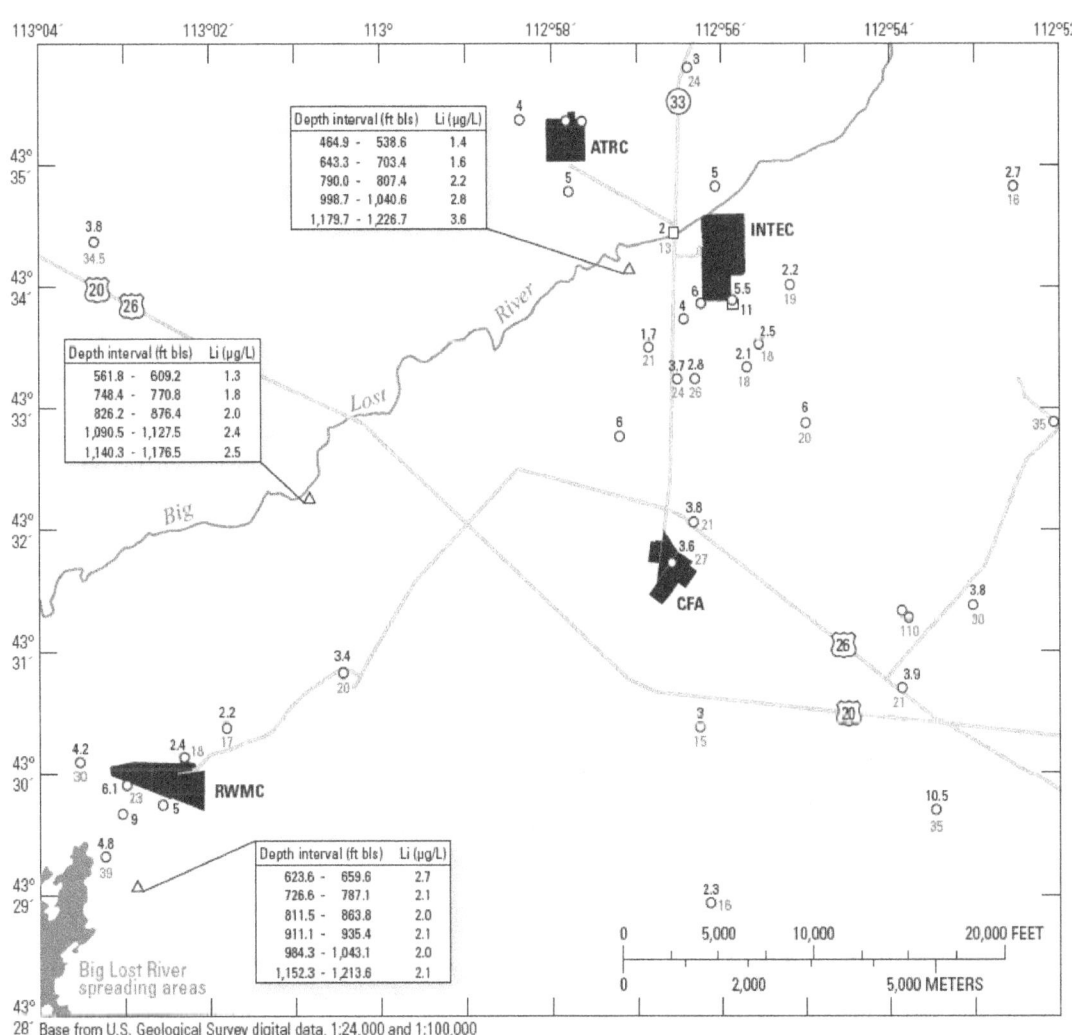

Depth interval (ft bls)	Li (µg/L)
464.9 - 538.6	1.4
643.3 - 703.4	1.6
790.0 - 807.4	2.2
998.7 - 1,040.6	2.8
1,179.7 - 1,226.7	3.6

Depth interval (ft bls)	Li (µg/L)
561.8 - 609.2	1.3
748.4 - 770.8	1.8
826.2 - 876.4	2.0
1,090.5 - 1,127.5	2.4
1,140.3 - 1,176.5	2.5

Depth interval (ft bls)	Li (µg/L)
623.6 - 659.6	2.7
726.6 - 787.1	2.1
811.5 - 863.8	2.0
911.1 - 935.4	2.1
984.3 - 1,043.1	2.0
1,152.3 - 1,213.6	2.1

Base from U.S. Geological Survey digital data, 1:24,000 and 1:100,000
Albers Equal-Area Conic projection, standard parallels 42°50'N, 44°10'N;
central meridian 113°00'W; North American Datum of 1927.

B. Southwest part of Idaho National Laboratory

EXPLANATION

Site facilities

ATRC—Advanced Test Reactor Complex
CFA—Central Facilities Area
INTEC—Idaho Nuclear Technology and Engineering Center
RWMC—Radioactive Waste Management Complex

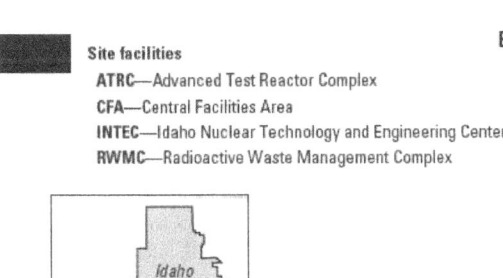

10.5 / 35   **Well sampled**—Entries 10.5 and 35 are the lithium (Li) and boron concentrations in micrograms per liter (µg/L), respectively. Average values used when wells sampled more than once.

2 / 13   **Surface-water or spring sample site**—Entries 2 and 13 are the lithium and boron concentrations in µg/L, respectively.

△   **Multilevel-monitoring well**—Depth interval in feet below land surface (ft bls)

Figure 7.—Continued

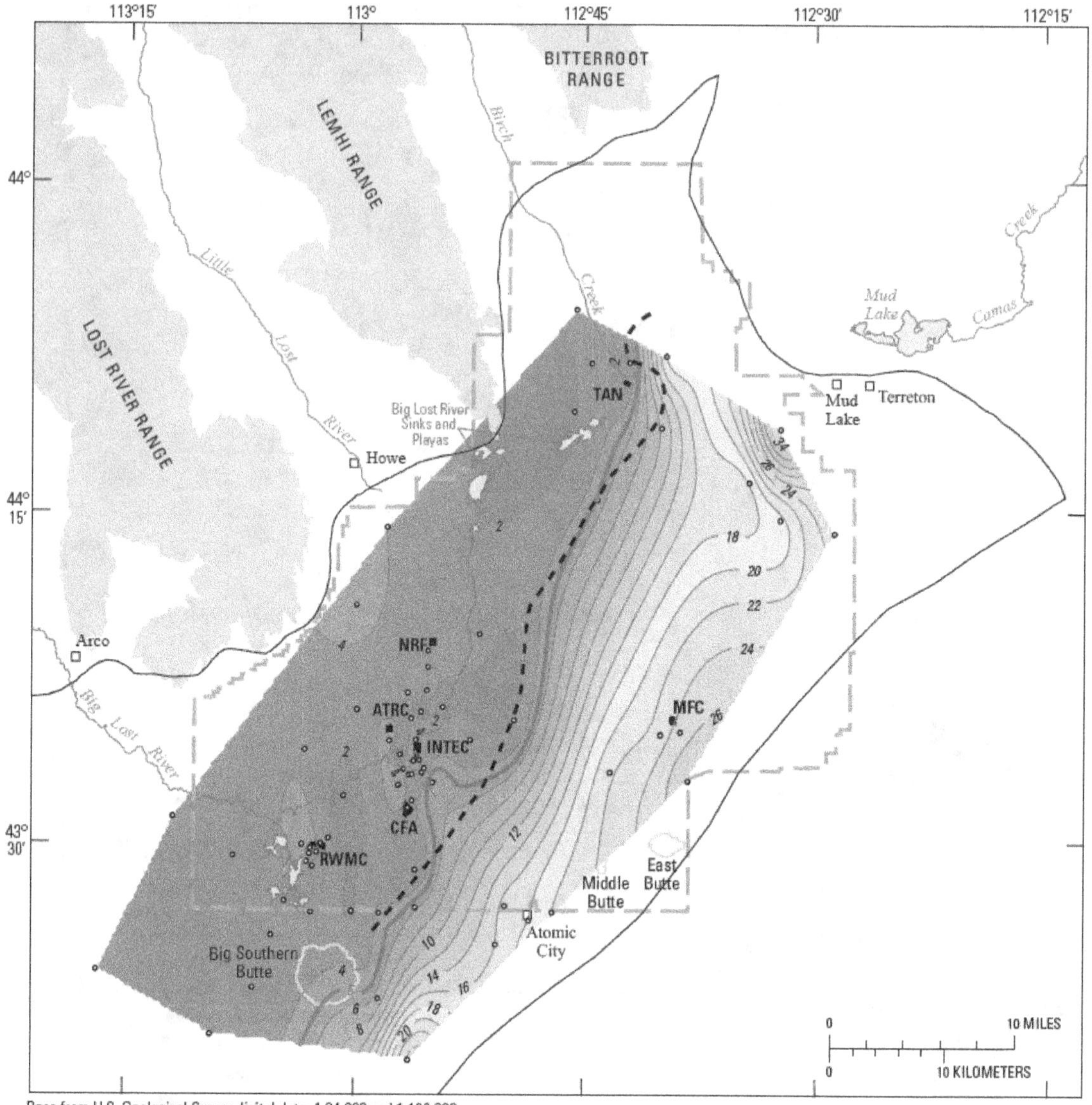

Figure 8.   Isopleths of lithium concentration in groundwater within the upper 200 feet of the eastern Snake River Plain aquifer, Idaho National Laboratory and vicinity, Idaho.

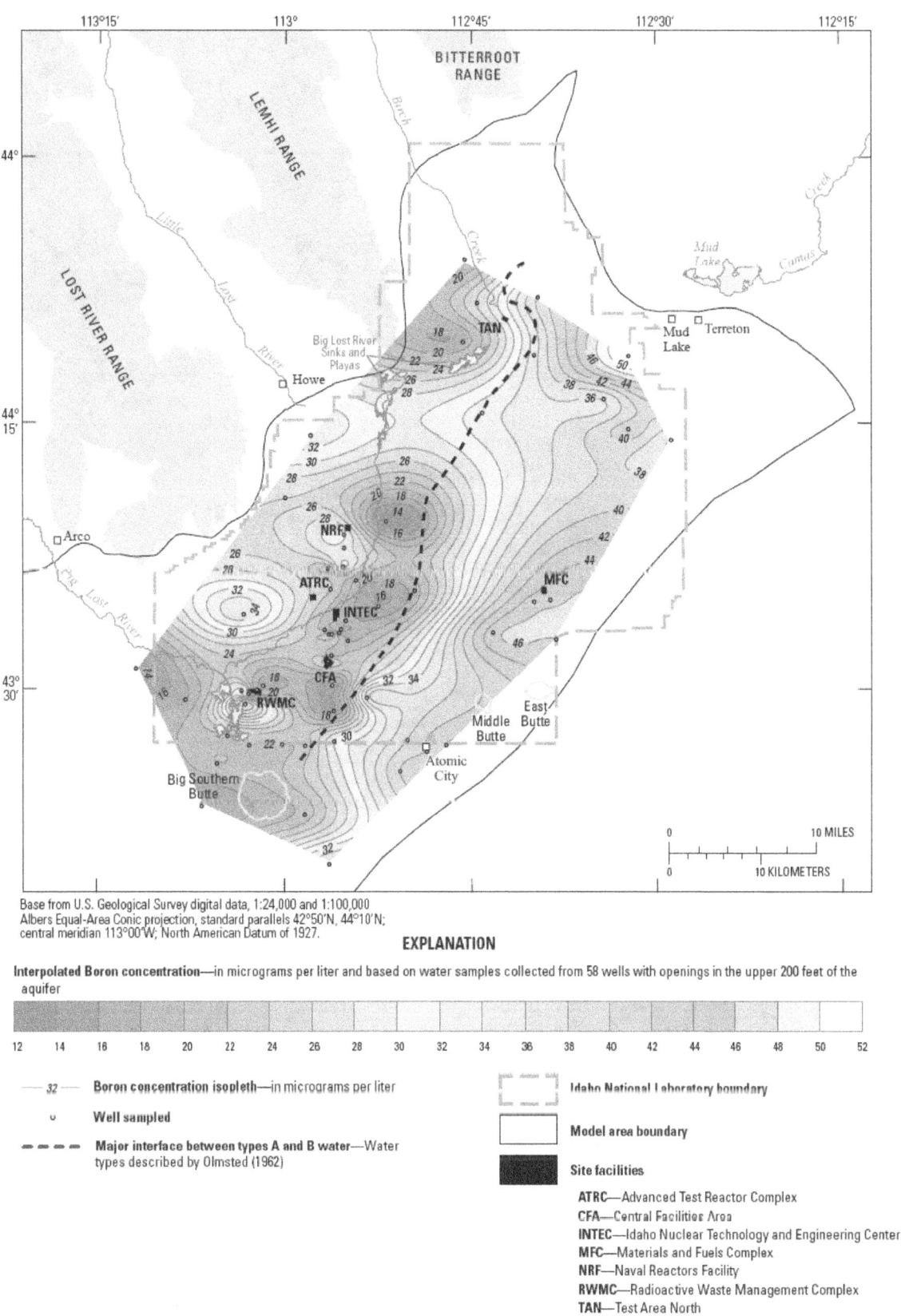

EXPLANATION

Interpolated Boron concentration—in micrograms per liter and based on water samples collected from 58 wells with openings in the upper 200 feet of the aquifer

| 12 | 14 | 16 | 18 | 20 | 22 | 24 | 26 | 28 | 30 | 32 | 34 | 36 | 38 | 40 | 42 | 44 | 46 | 48 | 50 | 52 |

— 32 — Boron concentration isopleth—in micrograms per liter

ᵒ Well sampled

▬ ▬ ▬ Major interface between types A and B water—Water types described by Olmsted (1962)

▭ Idaho National Laboratory boundary

▭ Model area boundary

▪ Site facilities

ATRC—Advanced Test Reactor Complex
CFA—Central Facilities Area
INTEC—Idaho Nuclear Technology and Engineering Center
MFC—Materials and Fuels Complex
NRF—Naval Reactors Facility
RWMC—Radioactive Waste Management Complex
TAN—Test Area North

Figure 9.   Isopleths of boron concentration in groundwater within the upper 200 feet of the eastern Snake River Plain aquifer, Idaho National Laboratory and vicinity, Idaho.

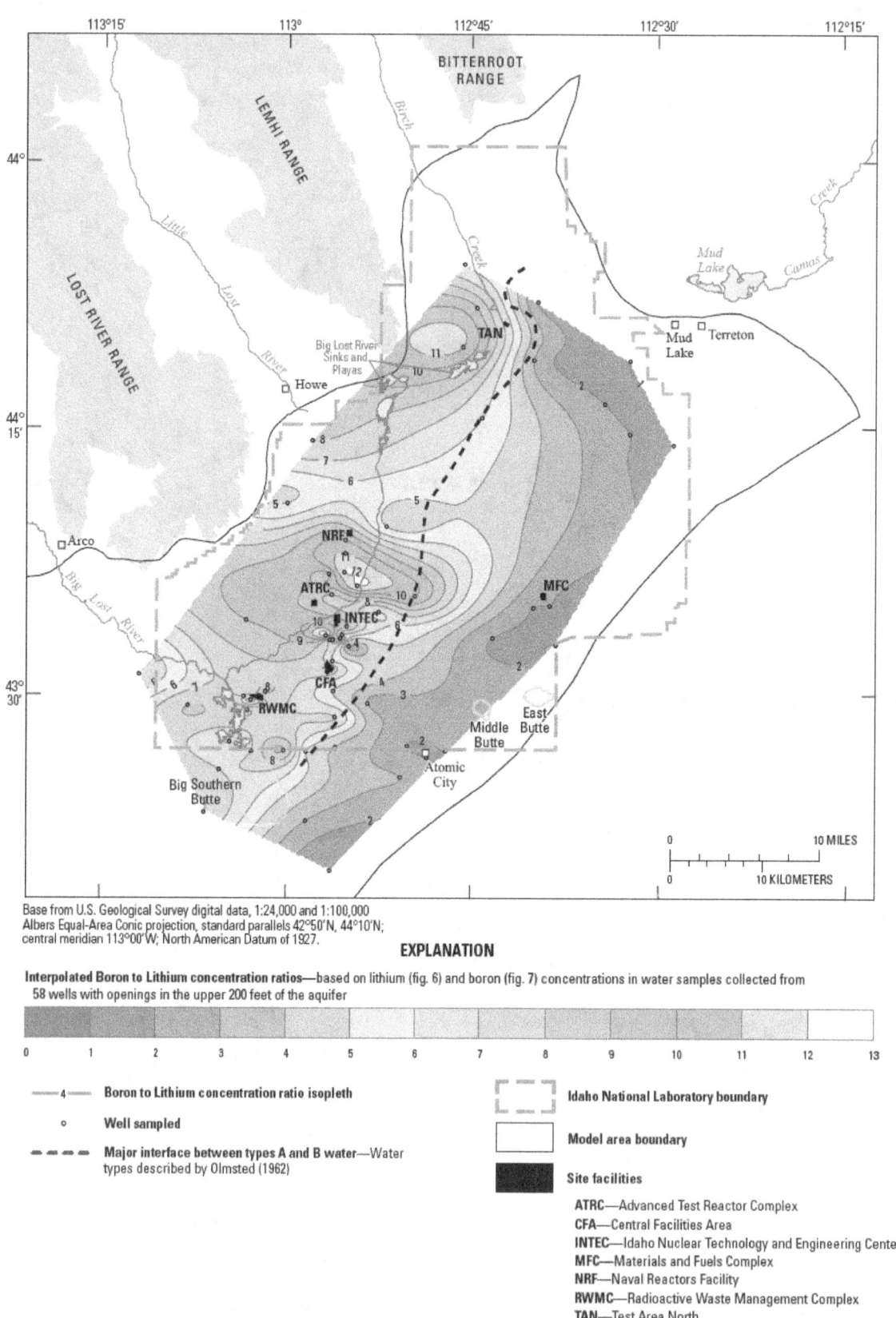

Base from U.S. Geological Survey digital data, 1:24,000 and 1:100,000
Albers Equal-Area Conic projection, standard parallels 42°50'N, 44°10'N;
central meridian 113°00'W; North American Datum of 1927.

EXPLANATION

Interpolated Boron to Lithium concentration ratios—based on lithium (fig. 6) and boron (fig. 7) concentrations in water samples collected from
58 wells with openings in the upper 200 feet of the aquifer

0   1   2   3   4   5   6   7   8   9   10   11   12   13

—— 4 ——   Boron to Lithium concentration ratio isopleth

○   Well sampled

– – – –   Major interface between types A and B water—Water
types described by Olmsted (1962)

Idaho National Laboratory boundary

Model area boundary

Site facilities

ATRC—Advanced Test Reactor Complex
CFA—Central Facilities Area
INTEC—Idaho Nuclear Technology and Engineering Center
MFC—Materials and Fuels Complex
NRF—Naval Reactors Facility
RWMC—Radioactive Waste Management Complex
TAN—Test Area North

Figure 10.   Isopleths of boron to lithium concentration ratio in groundwater within the upper 200 feet of the eastern
Snake River Plain aquifer, Idaho National Laboratory and vicinity, Idaho.

Boron concentrations and B/Li concentration ratios, ordered from lowest to highest and highest to lowest, respectively, also are shown in figure 11 for all well sites with a corresponding Li concentration measurement (n=87). The rank-ordered plot for B is similar to that for Li; however, unlike the Li plot, there is no sharp break in B concentrations. In general, B concentrations tend to increase from west to east with higher B concentrations (> 20 µg/L) and lower B/Li concentration ratios (< 6) east of the 5 µg/L Li line (figs. 9 and 10).

A scatter plot of Li versus B, and Li versus B/Li concentration ratios (n = 87 for each of these variables) indicates a well-defined region for groundwater west of the 5 µg/L Li line (figs. 12A and 8). In this region Li concentrations range from 1 to 4.8 µg/L, B concentrations

range from 11 to 39 µg/L, and B/Li ranges from 4 to 12 µg/L. This region is interpreted to represent groundwater that is derived from tributary valley underflow, streamflow-infiltration recharge, and agricultural return flow downgradient of the Howe agricultural area. A larger scale scatter plot of Li and B data for Li concentrations less than 5 µg/L (fig. 12B) indicates no strong correlation between Li and B concentrations (Spearman's rank correlation coefficient [ρ] = 0.56, the probability of an observed result arising by chance [p-value] = 3.4×10^{-5}, n=48, and slightly positive correlation) and between Li and B/Li concentration ratios (ρ=-0.63, p-value = 1.9×10^{-6}, and n=48; slightly negative correlation). Spatially, there appears to be no obvious trend or clustering of either Li or B concentrations (fig. 7).

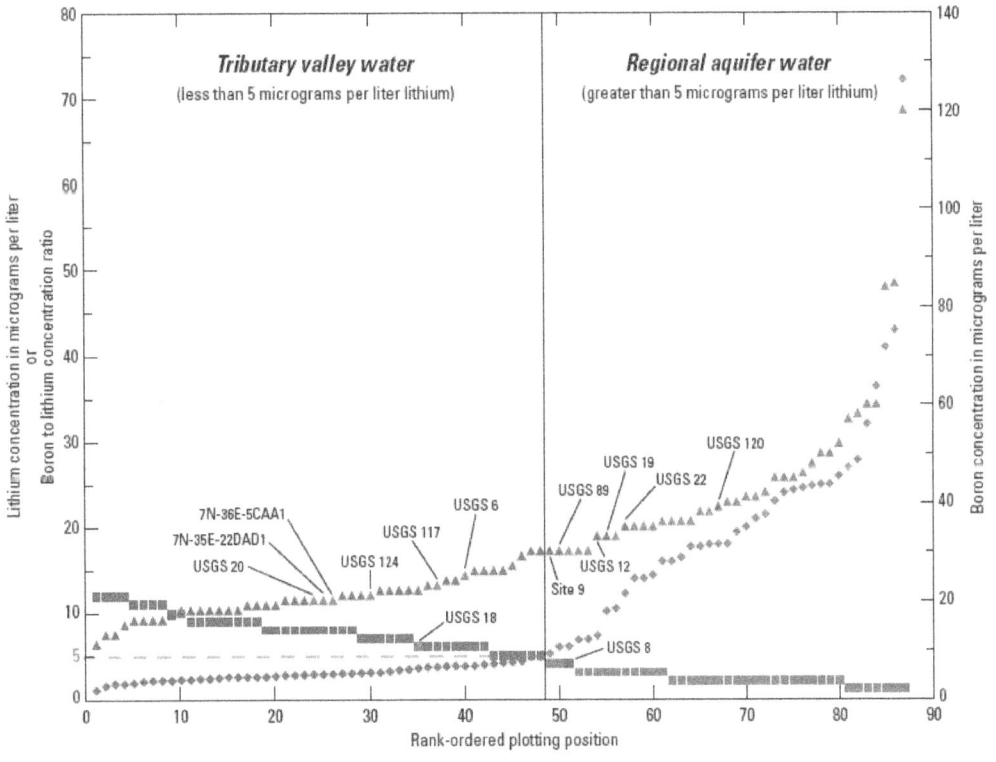

EXPLANATION

♦ **Lithium concentration**—measurements with lithium (Li) concentrations less than 5 micrograms per liter (µg/L) and ordered from smallest to largest (n = 48)

♦ **Lithium concentration**—measurements with Li concentration greater than 5 µg/L and ordered from smallest to largest (n = 39)

▲ **Boron concentration**—measurements with Li concentrations less than 5 µg/L and ordered from smallest to largest (n = 48)

▲ **Boron concentration**—measurements with Li concentrations greater than 5 µg/L and ordered from smallest to largest (n = 39)

■ **Boron to lithium concentration ratio**—measurements with Li concentrations less than 5 µg/L and ordered from largest to smallest (n = 48)

■ **Boron to lithium concentration ratio**—measurements with Li concentrations greater than 5 µg/L and ordered from largest to smallest (n = 39)

Figure 11. Rank-ordered plot of lithium concentrations, boron concentrations, and boron to lithium concentration ratios in groundwater from 87 wells, Idaho National Laboratory and vicinity, Idaho.

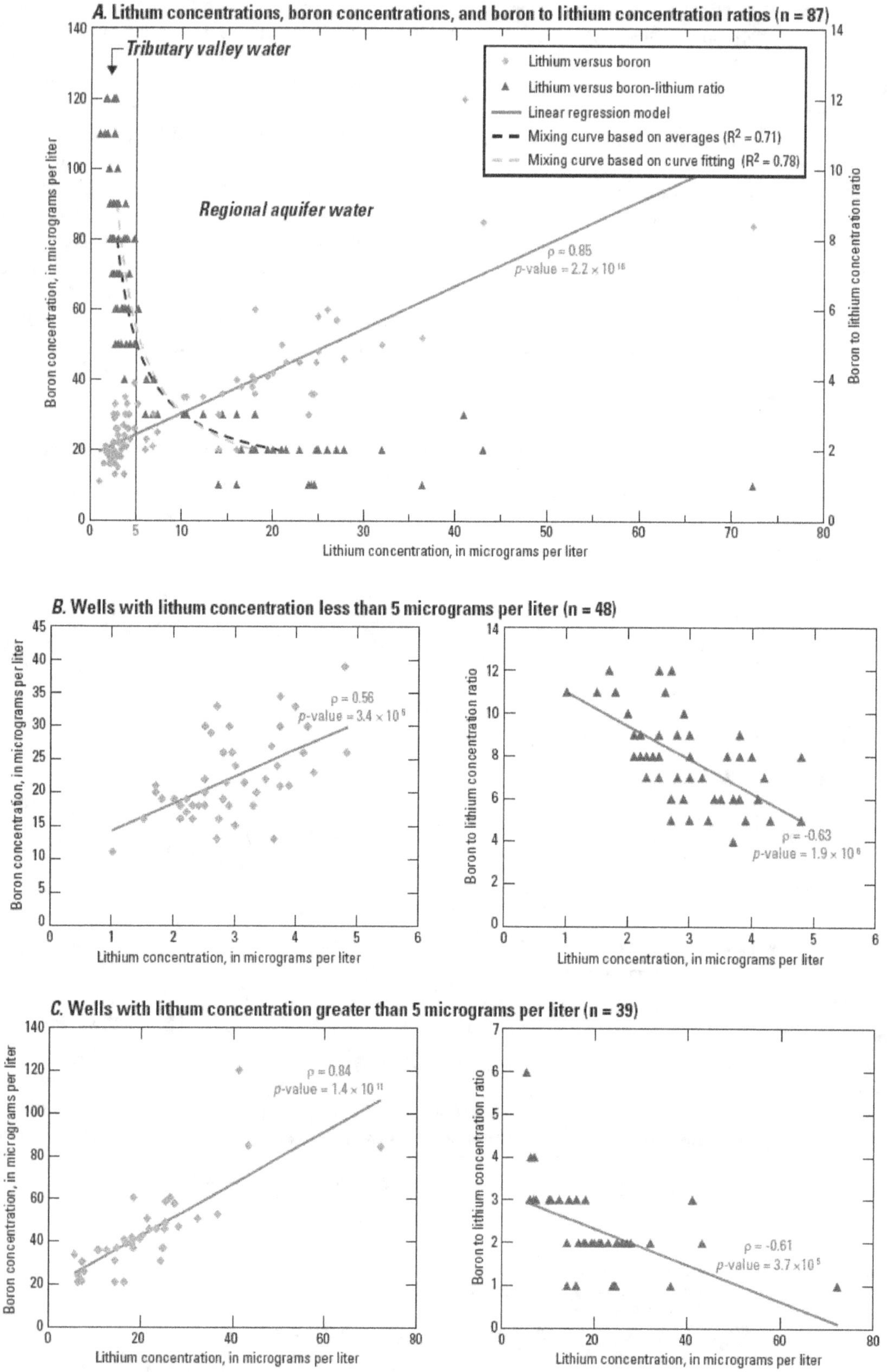

Figure 12. Lithium concentrations, boron concentrations, and boron to lithium concentration ratios in groundwater from (*A*) 87 wells in the study area, (*B*) 48 wells west of the 5-microgram-per-liter lithium line, and (*C*) 39 wells east of the 5-microgram-per-liter lithium line, Idaho National Laboratory and vicinity, Idaho. The location of wells is shown in figure 7 and the location of the 5-microgram-per-liter lithium line is shown in figure 8.

Although measurement sites were not selected randomly, the summary statistics (n=48) for Li, B, and B/Li west of the 5 µg/L Li line (table 3) indicate that each of these trace elements and their ratios are normally distributed with means, mediums, and modes that are almost identical. Under the assumption of normality, the standard deviation for Li concentrations indicates a well-defined mean or central tendency implying that 99.7 percent of the true population is within three standard deviations of the measured mean (2.9 ± 2.7 µg/L), a statistic that is well supported by the available data. Collectively the summary statistics characterize a Li population west of the 5 µg/L Li line that is statistically homogeneous. The single population outlier is the IET 1 Disposal well (figs. 1A and 7A) where the Li concentration is low (2.3 µg/L) and consistent with its geographic position west of the 5 µg/L Li line; however, its B concentration (72 µg/L) is much higher than other members of the population, resulting in a very high B/Li concentration ratio (31). The reason for the very high B concentration in this well is unclear but likely is the result of contamination from wastewater disposal (Busenberg and others, 2001, p. 51, "IET disposal water has been contaminated and could not be dated."). For this reason the IET disposal well has been excluded from the summary statistics shown in table 3.

Statistically, the background concentrations of Li and B in regional aquifer water, defined in this report as water east of the 5 µg/L line and consisting of water types described by Busenberg and others (2001), are much more heterogeneous than water derived from tributary valley underflow and streamflow-infiltration recharge west of the 5 µg/L Li line. The Li, B, and B/Li summary statistics (n=39) for groundwater east of the 5 µg/L Li line (table 3) indicate slightly skewed population distributions for Li and B and relatively large standard deviations with respect to the mean values for these two trace elements. East of the 5 µg/L Li line, Li concentrations range from 5.2 to 72.3 µg/L with a mean of 20.8 µg/L and a standard deviation of 12.3 µg/L; and B concentrations range from 20 to 120 µg/L with a mean of 43 µg/L and a standard deviation of 19 µg/L. The B/Li concentration ratio is well defined with a small standard

deviation and a mean, medium, and mode that are identical, suggesting a normal distribution. The chemical or physical basis for the well-defined B/Li concentration ratio east of the 5 µg/L Li line (table 3) is not known. The observation is presented simply to draw attention to a population characteristic of regional aquifer water that distinguishes it from tributary valley water. The standard deviation for B/Li (1) indicates that more than 95 percent of the true population is within two standard deviations of the measured mean ratio (2±2), again a statistic that is well supported by the range of measured values, 1 to 6. A larger scale scatter plot of Li and B data for Li concentrations greater than 5 µg/L (fig. 12C) indicates a moderately positive correlation between Li and B concentrations ($\rho=0.84$, p-value $= 1.4\times10^{-11}$, and n=39) and a slightly negative correlation between Li and B/Li concentration ratios ($\rho=-0.61$, p-value $= 3.7\times10^{-5}$, and n = 39).

Trends in the rank-ordered distribution of Li, B, and B/Li (fig. 11) and differences in the scatter plots of these data west and east of the 5 µg/L Li line (figs. 12A, 12B, 12C) characterize a region of the aquifer where the well-defined statistical characteristics of these trace elements in tributary valley groundwater abruptly give way to a more diverse population that is dominated by regional aquifer water. Groundwater west of the 5 µg/L Li line represents tributary valley underflow that is mixed with recharge across the water table boundary primarily from streamflow in the Big Lost River, Little Lost River, and Birch Creek with the largest contributions coming from the Big Lost River channel, sinks, playas, and spreading areas (Ackerman and others, 2010, table 4). The heterogeneity and higher background concentrations of Li and B in regional aquifer water east of the 5 µg/L Li line likely are the result of spatially variable dissolution and interaction of groundwater with silicic volcanic rocks and buried thermal spring deposits, mixing of groundwater with locally elevated inputs of Li and B from active thermal springs and subsurface geothermal sources, and dilution of groundwater from surface-water sources with low Li and B concentrations.

Table 3. Summary statistics for paired boron and lithium concentrations in groundwater at the Idaho National Laboratory and vicinity, Idaho.

[Location of 5 microgram per liter lithium line is shown in figure 8. Lithium and boron: concentrations in microgram per liter. The IET 1 Disposal well (435153112420501) was excluded from the analysis as an outlier. **Abbreviations:** µg/L, microgram per liter; Li, lithium; NA, not applicable]

Groundwater region with respect to 5 µg/L Li line	Number of samples	Lithium					Boron					Boron / Lithium				
		Range	Mean	Median	Mode	Standard deviation	Range	Mean	Median	Mode	Standard deviation	Range	Mean	Median	Mode	Standard deviation
West	48	1–4.8	2.9	2.8	2.5	0.9	11–39	22	21	18	6	4–12	8	8	8	2
East	39	5.2–72.3	20.8	18.0	18.0	12.3	20–120	43	40	36	19	1–6	2	2	2	1

Along the northeast model boundary, regional aquifer water includes contributions of water from local sources with low Li and B concentrations relative to regional mean values. These sources include streamflow-infiltration recharge from Camas Creek and Mud Lake and surface-water irrigation return flow in the Mud Lake and Terreton areas. Groundwater from wells in this area (figs. 1A and 4, table 4) was classified as Type III water by Busenberg (2001, p. 68). In the area between the MFC and Atomic City, rapid focused recharge in closed drainage basins provides a source of water with low Li and B concentrations. Groundwater from wells in this area (figs. 1A and 4, table 4) was classified as Type II water by Busenberg (2001, p. 68). To some extent the dilution effect of these local contributions can produce a mixture of regional aquifer water that is nearly indistinguishable from regional aquifer water that has been mixed with tributary valley groundwater.

Derivation of a Mixing Model for Tributary Valley Water and Regional Aquifer Water

A two-component mixing model, using Li concentrations and B/Li concentration ratios, was developed to describe a mixing zone where inflow (regional aquifer water) along the northeast model boundary mixes with inflow (tributary valley groundwater) along the northwest mountain-front boundary (appendix F). The general mixing equation for Li and B/Li variables is a hyperbola of the form:

$$B[\text{Li}]_{mix}\left(\text{B/Li}\right)_{mix} + C[\text{Li}]_{mix} + D = 0 \qquad (1)$$

where

$[\text{Li}]_{mix}$ is the Li concentration of water measured in the mixing zone in µg/L,

$\left(\text{B/Li}\right)_{mix}$ is the B/Li concentration ratio of water measured in the mixing zone, and

B, C, D are coefficients of end-member variables Li and B/Li.

In element-ratio space the coefficients of equation (1) are given as

$$B = [\text{Li}]_{ra} - [\text{Li}]_{tv}$$
$$C = [\text{Li}]_{tv}\left(\text{B/Li}\right)_{tv} - [\text{Li}]_{ra}\left(\text{B/Li}\right)_{ra}$$
$$D = [\text{Li}]_{tv}[\text{Li}]_{ra}\left(\text{B/Li}\right)_{ra} - [\text{Li}]_{ra}[\text{Li}]_{tv}\left(\text{B/Li}\right)_{tv} \qquad (2)$$

where

$[\text{Li}]_{ra}$ is the Li concentration of regional water in µg/L,

$[\text{Li}]_{tv}$ is the Li concentration of tributary valley water in µg/L,

$\left(B/Li\right)_{ra}$ is the B/Li concentration ratio of regional aquifer water, and

$\left(B/Li\right)_{tv}$ is the B/Li concentration ratio of tributary valley water.

Mean values of Li concentrations and B/Li concentration ratios were used to represent the end-member contributions from (1) tributary valley groundwater west of the 5 µg/L Li line, and (2) regional aquifer water east of the 5 µg/L Li line; and to compute the coefficients of the hyperbolic mixing equation (2). Using mean values, the mixed zone is bounded by Li concentrations of 2.9 µg/L (standard deviation [σ]=0.9) for tributary valley water and 20.8 µg/L (σ = 12.3) for regional aquifer water, and B/Li concentration ratios of 2 (σ=1) for regional aquifer water and 8 (σ=2) for tributary valley water. Substituting mean values into equation (2) gives the general mixing model (equation 1) as

$$17.9[\text{Li}]_{mix}\left(\text{B/Li}\right)_{mix} - 18[\text{Li}]_{mix} - 362 = 0$$

Although the model fit to the Li and B/Li data is relatively good (coefficient of determination [R^2] =0.71; fig. 12A and appendix F), the reliability of the end-member estimate for the Li concentration of regional aquifer water is very uncertain. This uncertainty is reflected in the range (5.2 to 72.3 µg/L) and large standard deviation (12.3 µg/L) for Li concentrations from regional aquifer groundwater relative to its mean value (20.8 µg/L), and in the strong northeast to southwest spatial trend in Li isopleths east of the 5 µg/L Li line (fig. 8).

An alternative two-component mixing model was developed using a curve-fitting regression technique to identify end-member variables in equation (2) that optimize the fit of the hyperbolic equation (1) to the observed [Li] and (B/Li). A unique solution to the optimization problem required two of the four end-member variables be held constant. Variables with small standard deviations, the Li concentration of tributary valley water and the B/Li concentration ratio of regional aquifer water, were specified using their population means, and those variables with large uncertainty, the Li concentration of the regional water and the B/Li concentration ratio of the tributary valley water, were estimated in the regression, independent of their population means. The resulting regressed mixing model is given as:

$$10.9[\text{Li}]_{mix}\left(\text{B/Li}\right)_{mix} - 11[\text{Li}]_{mix} - 192 = 0 \qquad (4)$$

The regression-based model, using average values for $[Li]_{tv}$ and $(B/Li)_{ra}$ to compute $[\text{Li}]_{ra}$ and $(\text{Bi/Li})_{tv}$, defines the boundaries of the mixing zone as:

Model 1:

2.9 µg/L $\leq [\text{Li}]_{mix} \leq 18$ µg/L and $2 \leq (\text{B/Li})_{mix} \leq 7$,

where

$[\text{Li}]_{tv} = [Li]_{tv} = 2.9$ µg/L and $(\text{B/Li})_{ra} = (B/Li)_{ra} = 2.$

Table 4. Chemical characteristics of groundwater in the vicinity of and east of the 5-microgram-per-liter lithium line.

[**Regression-based model:** OB, outside bounds of all regression-based models. **Abbreviations:** µg/L, microgram per liter; –, not available; NA, not applicable]

Local name	Open interval in feet below water table	Pump intake in feet below water table	Model layer	Lithium (µg/L)	Boron (µg/L)	Boron / lithium ratio	Regression-based model
Regional groundwater mixed with rapid focused recharge (Type II water of Busenberg, 2001)							
Arbor Test 1	0–48	38	1	24.7	44	2	NA
Area II	3–50 / 181–203	30	1, 2	17.7	41	2	NA
Atomic City	0–52	28	1	18.0	40	2	NA
Leo Rogers 1	0–89	–	1	16.0	40	3	NA
USGS 1	10–20	22	1	18.0	42	2	NA
USGS 2	13–33	21	1	20.4	45	2	NA
USGS 14	3–6	22	1	24.3	36	1	NA
USGS 100	0–72	18	1	23.4	44	2	NA
USGS 101	0–92	17	1	27.7	45	2	NA
USGS 110A	0–91	46	1	15.9	38	2	NA
Regional groundwater mixed with irrigation return flow, Camas Creek, and (or) Mud Lake water (Type III water of Busenberg, 2001)							
Engberson	–	–	1	14.4	36	3	NA
USGS 27	21–31 / 69–79	33	1	36.4	52	1	NA
USGS 32	12–30 / 30–98	28	1	19.1	43	2	NA
USGS 29	4–39 / 39–66	43	1	23.7	36	2	NA
USGS 31	26–46 / 47–169	25	1, 2	17.8	35	2	NA
USGS 4	22–52 / 59–290	40	1, 2, 3	24.2	48	2	NA
Regional groundwater, tributary valley water, streamflow infiltration recharge or mixtures of these							
ANP 9	10–87	43	1	10.2	35	3	1, 2
NPR Test[1]	33–69	20	1	2.0	16	8	OB
USGS 5[1]	4–26	17	1	2.0	19	10	OB
USGS 18	23–47	26	1	5.2	33	6	1, 2, 3, 4
USGS 26[2]	18–52	41	1	18.4	38	2	OB
BFW	43–154	–	1, 2	3.9	21	5	1, 2, 3, 4
USGS 103	0–174	115	1, 2	6.9	29	4	1, 2, 3, 4
USGS 104[1]	0–142	34	1, 2	2.4	16	7	OB
USGS 107	0–209	49	1, 2	10.5	35	3	1, 2
USGS 124	66–115	52	1, 2	6.9	20	3	1, 2, 3, 4
USGS 6	115–200	40	2	7.3	25	3	1, 2, 3, 4
Site 9[1]	206–580	49	3, 4, 5	3.5	30	9	OB
Site 14[2]	259–439	49	3, 4	12.3	35	3	OB
USGS 7[2]	239–259 / 760–1,200	242	3, 4, 5	25.9	57	2	OB

[1]Interpreted as tributary valley water
[2]Interpreted as regional aquifer water

The average values used for the end members $[Li]_{tv} = 2.9$ µg/L and $(B/Li)_{ra} = 2$, although based on well defined statistical characteristics, are approximate. To precisely define the boundaries of the mixing zone all four end members should be single valued and representative of the contributing sources immediately prior to the onset of mixing. This constraint requires the contributing sources to be well mixed, which is clearly not the case, particularly for [Li] concentrations east of the 5 µg/L Li line and (B/Li) concentration ratios throughout the model area. Contour maps of [Li] and (B/Li) (figs. 8 and 10) provide a spatial frame of reference that can be used to refine the estimates for $[Li]_{tv}$ and $(B/Li)_{ra}$ used in Model 1. These refinements are based on a comparison of average estimates to the spatial distribution of measured values east and west of the 5 µg/L Li line. Refinements using this approach are subjective; however, these refinements derive support from [Li] and (B/Li) spatial trends that parallel the regional direction of groundwater flow and from the water-typing classifications of Olmsted and Busenberg.

Sensitivity analyses (appendix F), where the average values of $[Li]_{tv}$ and $(B/Li)_{ra}$ are increased by 1σ, provide three alternative models that define the boundaries of the mixing zone as:

Model 2:

3.8 µg/L $\leq [Li]_{mix} \leq 18$ µg/L and $2 \leq (B/Li)_{mix} \leq 6$,

where

$[Li]_{tv} = [Li]_{tv} + 1\sigma = 3.8$ µg/L and $(B/Li)_{ra} = (B/Li)_{ra} = 2$;

Model 3:

2.9 µg/L $\leq [Li]_{mix} \leq 9$ µg/L and $3 \leq (B/Li)_{mix} \leq 7$,

where

$[Li]_{tv} = [Li]_{tv} = 2.9$ µg/L and $(B/Li)_{ra} = (B/Li)_{ra} + 1\sigma = 3$;

Model 4:

3.8 µg/L $\leq [Li]_{mix} \leq 9$ µg/L and $3 \leq (B/Li)_{mix} \leq 6$,

where

$[Li]_{tv} = [Li]_{tv} + 1\sigma = 3.8$ µg/L and $(B/Li)_{ra} = (B/Li)_{ra} + 1\sigma = 3$;

The curve-fitting statistics for all four regression-based models are identical with $R^2 = 0.78$ and the RMSE = 1.47. These statistics indicate that all four models produce equally valid results even though the models are not mathematically equivalent. Model 1 imposes the least restrictive bounds on the $[Li]_{tv}$ and $(B/Li)_{ra}$ end members; Model 4 imposes the most restrictive bounds on these end members.

Character of the Mixing Zone

The contour map of [Li] (fig. 8) indicates dominant northeast to southwest linear trends beginning with concentrations greater than about 4 µg/L Li, and reasonably well-mixed water west of this line suggests that an end-member estimate for $[Li]_{tv} = 3.8$ µg/L is probably more representative than a mean value of 2.9 µg/L (Models 2 and 4). The contour map of (B/Li) (fig. 10) indicates that values of $(B/Li)_{ra} = 2$ are very distant from the 5 µg/L Li line, suggesting that a value of $(B/Li)_{ra} = 3$ is probably more representative than a mean value of 2 (Models 3 and 4).

Wells in the vicinity of and east of the 5 µg/L Li line that most closely approximate the bounding criteria for mixed water using the regression-based models are summarized in table 4 and shown in figure 13. Those wells with the Type II and Type III waters defined by Busenberg (equivalent to the Type B water of Olmsted) are outside the bounds of Models 3 and 4. Several wells meet the criteria for Models 1 and 2. However, as previously discussed, Type II and Type III water is regional aquifer water that is mixed with water from local sources with low Li and B concentrations. The dilution effect of these local contributions can produce a mixture of regional aquifer water that is nearly indistinguishable from regional aquifer water that has been mixed with tributary valley groundwater.

The Li concentrations and B/Li concentration ratios for five wells with open intervals less than 200 ft below the water table (USGS 6, 18, 103, 124, and BFW) meet the criteria for all four models. Wells ANP 9 and USGS 107 meet the criteria for Models 1 and 2 and very nearly meet the criteria for Models 3 and 4 (10.2 and 10.5 µg/L Li versus 9 µg/L Li, respectively). Groundwater from the other wells with open intervals less than 200 ft below the water table, although in close proximity to the 5 µg/L Li line, fall outside of the mixing zone criteria and are interpreted as either tributary valley water (NPR Test, USGS 5, USGS 104, and Site 9) or regional aquifer water (USGS 26). The location and distribution of these wells suggest that the width of the mixing zone, within the upper 200 ft of the aquifer, is relatively narrow and is probably no more than about 1 to 2 mi wide along its dominant northeast to southwest trend (fig. 13).

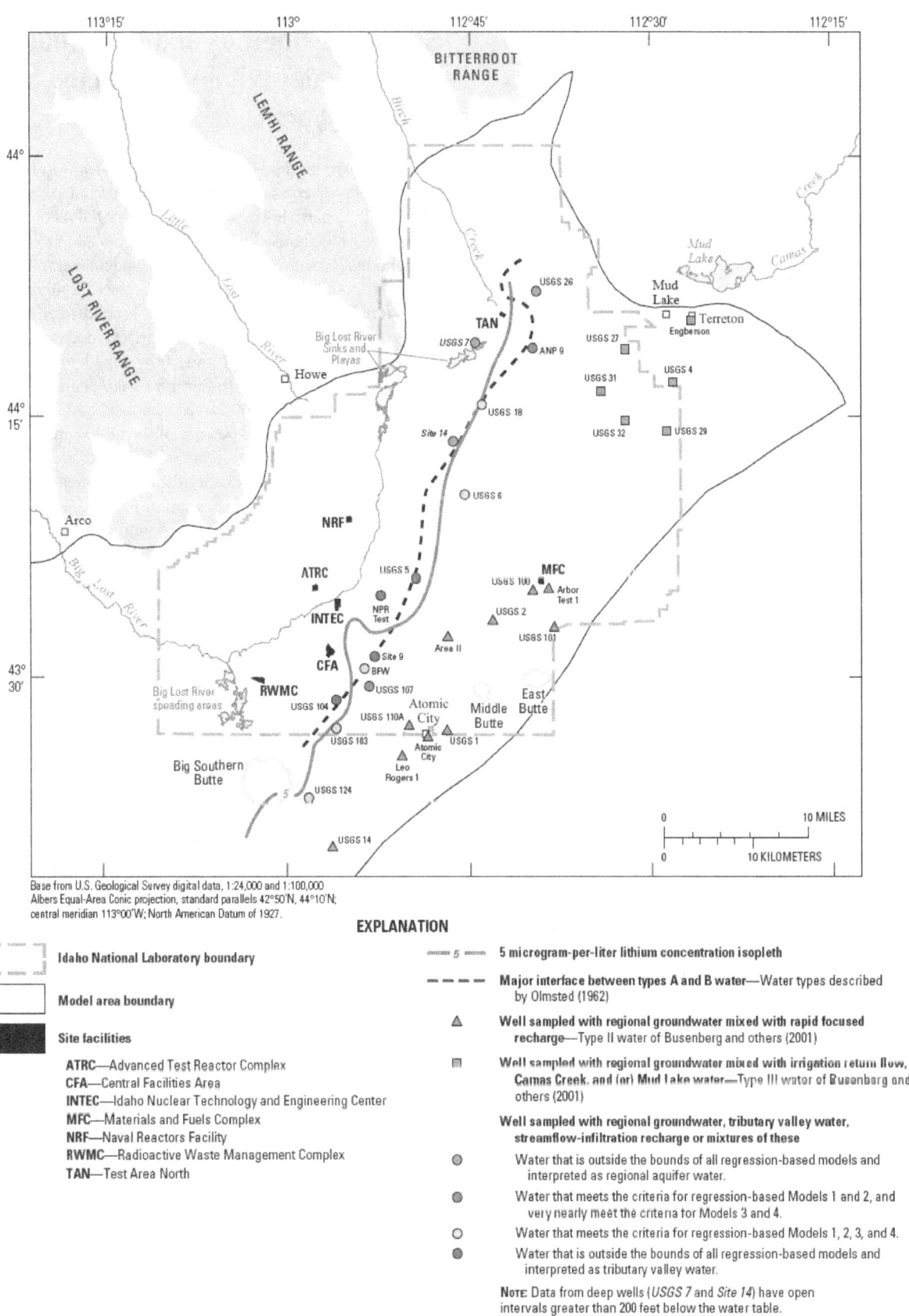

Base from U.S. Geological Survey digital data, 1:24,000 and 1:100,000
Albers Equal-Area Conic projection, standard parallels 42°50'N, 44°10'N;
central meridian 113°00'W; North American Datum of 1927.

EXPLANATION

Idaho National Laboratory boundary

Model area boundary

Site facilities

ATRC—Advanced Test Reactor Complex
CFA—Central Facilities Area
INTEC—Idaho Nuclear Technology and Engineering Center
MFC—Materials and Fuels Complex
NRF—Naval Reactors Facility
RWMC—Radioactive Waste Management Complex
TAN—Test Area North

5 microgram-per-liter lithium concentration isopleth

Major interface between types A and B water—Water types described by Olmsted (1962)

△ Well sampled with regional groundwater mixed with rapid focused recharge—Type II water of Busenberg and others (2001)

▣ Well sampled with regional groundwater mixed with irrigation return flow, Camas Creek, and (or) Mud Lake water—Type III water of Busenberg and others (2001)

Well sampled with regional groundwater, tributary valley water, streamflow-infiltration recharge or mixtures of these

○ Water that is outside the bounds of all regression-based models and interpreted as regional aquifer water.

● Water that meets the criteria for regression-based Models 1 and 2, and very nearly meet the criteria for Models 3 and 4.

○ Water that meets the criteria for regression-based Models 1, 2, 3, and 4.

● Water that is outside the bounds of all regression-based models and interpreted as tributary valley water.

NOTE: Data from deep wells (*USGS 7* and *Site 14*) have open intervals greater than 200 feet below the water table.

Figure 13.   Location of wells in the vicinity of and east of the 5-microgram-per-liter lithium line that most closely approximate the bounding criteria for mixed water using the regression-based general mixing model.

Data from a limited number of deeper wells in the northern and northwestern part of the model area and west of the 5 µg/L Li line (USGS 7 and Site 14) suggest that the boundary separating tributary valley water from regional aquifer water probably shifts westward with increasing depth in the aquifer forming a wedge of predominantly tributary valley underflow and streamflow-infiltration recharge from Birch Creek that displaces and overlies regional aquifer underflow along the northeast model boundary.

East of the CFA and along the southern end of the line separating Type A and Type B water the close proximity of Type A water to Type B water in both shallow and deep wells and the abrupt change in Li concentrations across the 5 µg/L Li line suggests that the mixing zone separating tributary valley water from regional aquifer water also is narrow at this location and may persist to a considerable depth.

The character of the mixing zone south of the CFA and close to the southern boundary of the INL is indicated by data from well USGS 103, a 1,307-ft-deep well that was instrumented with MLMS packers in 2007 that isolate seven sample zones from about 85 to 695 ft below the water table (Bartholomay and Twining, 2010). The upper zone of water (85 to 107 ft below the water table) has a silica concentration representative of precipitation recharge; the Li concentration greater than 5 µg/L, the sodium and potassium ratio of 24 percent cations (appendix D), and the fluoride concentration of 0.38 mg/L are representative of northeast regional recharge (Type B) indicating that this aquifer zone may be a mixture of predominantly Type B water with some Type A and (or) precipitation recharge. Zone 5 (about 307 to 335 ft below the water table) and zone 6 (about 182 to 247 ft below the water table) probably represent a mixture of Type A and Type B water that is slightly more dominated by Type B water as evidenced by decreasing sodium and potassium ratios (but still greater than 15 percent) and decreasing fluoride concentrations from zone 7 (appendix D). Type A water is predominant in the four lower zones (appendix D) and tritium concentrations indicate some recharge from wastewater disposal. The deeper occurrence of Type A water in this area may be evidence of the displacement of slower moving tributary valley water by faster moving regional water in the upper part of the aquifer.

Simulated Sources and Velocities of Groundwater within Each of the Six Model Layers

Backward particle tracking, using the steady-state flow model (Ackerman and others, 2010) and modified version of the particle-tracking program MODPATH (appendix A), were used to simulate sources and velocities of groundwater within each of the six model layers. Particles were distributed uniformly in model layers 1 through 6 in a layered-grid configuration to track groundwater flow back to inflow sources across the northwest mountain-front boundary, northeast regional-underflow boundary, and water table boundary (fig. 3). Particles were distributed over (1) 100 percent of the model area in layers 1 and 2 (7,763 particles per layer), (2) 98 percent of the model area in layer 3 (7,626 particles), (3) 96 percent of the model area in layer 4 (7,480 particles), (4) 78 percent of the model area in layer 5 (6,125 particles), and (5) 71 percent of the model area in layer 6 (5,541 particles) (table 5, appendix G). Differences in the number of particles released in each model layer reflect changes in the thickness of the aquifer and hence the number of active cells within each model layer (fig. 2). In each model layer one particle was released at the center of each 2 by 2 block of cells resulting in a particle density of one particle for every 0.25 mi^2. The backward particle-tracking simulation was run until all particles either (1) entered a specified flow cell along the northwest mountain-front boundary or northeast regional-underflow boundary (fig. 3) or (2) terminated in a weak source cell in which recharge from internal sources was greater than half of the total outflow from the cell (FRAC=0.5).

The simulated groundwater source area for a particle was based on its location at termination. For example, particles terminating in specified flow cells representing underflow from the Big Lost River valley were assigned to the Big Lost River valley source area. Specified source areas include (1) the Big Lost River (BLR), Little Lost River (LLR), and Birch Creek (BC) valleys, (2) the Reno (Re), Monteview (Mo), Mud Lake (ML), and Terreton (Te) sections of the northeast regional-underflow boundary, and (3) Big Lost River stream

Table 5.   Summary of backward particle-tracking results showing the percentage of source area contributions of tributary valley, streamflow-infiltration recharge, and regional aquifer water for particles released in a layered-grid configuration within the model domain, Idaho National Laboratory and vicinity, Idaho.

[Location of source areas are shown in figure 3. Orphans are identified as particles that did not terminate in one of the specified source areas. **FRAC:** the fraction of the total outflow from a cell contributed by an internal source; used to identify weak-source cells. **Layer:** the model layer particles are located in at their release. **Base case:** FRAC = 0.5; Layer = 1–6. **Abbreviation:** NA, not applicable]

Simulated groundwater source area	Percentage of total inflow from model boundaries	Base case	FRAC		Layer					
			0.1	0.9	1	2	3	4	5	6
Northwest mountain-front boundary										
Big Lost River valley (BLR)	17.1	12.6	11.6	12.6	15.6	16.5	15.3	14.9	7.2	2.3
Little Lost River valley (LLR)	10.5	14.0	10.3	16.3	18.3	20.0	19.6	11.6	6.1	3.8
Birch Creek valley (BC)	2.9	6.3	6.3	6.5	2.6	3.7	5.2	10.3	6.1	11.4
Northwest mountain-front subtotal	**30.5**	**32.9**	**28.2**	**35.4**	**36.5**	**40.3**	**40.2**	**36.8**	**19.4**	**17.5**
Northeast regional-underflow boundary										
Reno Ranch section (Re)	3.6	5.3	5.3	5.3	1.6	2.7	3.1	4.1	8.1	15.2
Monteview section (Mo)	4.8	5.1	5.1	5.1	2.4	3.3	3.1	4.7	7.9	11.6
Mud Lake section (ML)	16.9	18.1	18.1	18.1	17.8	16.9	21.7	24.2	16.8	8.1
Terreton section (Te)	32.5	31.4	31.4	31.4	22.5	26.2	25.8	27.7	46.3	47.4
Northeast regional-underflow subtotal	**57.9**	**59.9**	**59.9**	**59.9**	**44.4**	**49.1**	**53.8**	**60.8**	**79.2**	**82.4**
Water table boundary										
Streamflow infiltration										
Big Lost River infiltration										
Stream reach 600–601	1.0	0.1	2.1	0.0	0.2	0.2	0.1	0.1	0.0	0.0
Stream reach 602–605	2.0	0.0	1.6	0.0	0.0	0.0	0.0	0.0	0.0	0.0
Stream reach 606–607	1.2	0.8	1.3	0.5	3.9	0.1	0.0	0.1	0.0	0.0
Stream reach 608–610	1.7	2.8	3.5	0.5	6.0	5.7	3.0	0.6	0.3	0.0
Little Lost River infiltration										
Stream reach 611	0.1	0.0	0.2	0.0	0.0	0.0	0.0	0.0	0.0	0.0
Birch Creek infiltration										
Stream reach 612	0.9	2.2	2.0	2.2	3.5	3.8	2.6	1.4	0.8	0.1
Streamflow-infiltration subtotal	**7.0**	**5.9**	**10.7**	**3.2**	**13.6**	**9.8**	**5.8**	**2.3**	**1.1**	**0.1**
Orphans	**NA**	**1.3**	**1.2**	**1.5**	**5.5**	**0.8**	**0.3**	**0.1**	**0.3**	**0.0**

reaches 600–601, 602–605 (spreading areas), 606–607, and 608–610 (sinks and playas), Little Lost River stream reach 611, and Birch Creek stream reach 612 (fig. 3). Precipitation recharge, irrigation infiltration, and industrial water-use returns, representing 3.3, 1.0, and 0.3 percent of the total inflow, respectively, (table 1, figs. 3 and 4) were not included as sources because of their relatively small contribution to total inflow and their low flux densities relative to those of other contributing sources. Particles that did not terminate in one of the specified source areas are identified as "orphans." In general, orphans either terminated at the water table boundary in the vicinity of the Birch Creek stream or along the northwest mountain-front boundary.

Six Model Layer Simulation Results

Backward particle-tracking results for the base-case simulation (FRAC = 0.5) (table 5; appendix G) indicate that tributary valley underflows from the Big Lost River, Little Lost River, and Birch Creek valley source areas represent 12.6, 14.0, and 6.3 percent of the inflow sources, inflow from the northeast boundary represents 59.9 percent, and streamflow-infiltration recharge represents 5.9 percent. Orphans account for 1.3 percent of the total number of particles released in model layers 1 through 6. A comparison between source area contributions calculated using backward particle tracking and those determined from the boundary conditions of the steady-state flow model shows good agreement ($R^2 = 0.95$), with a maximum difference of 2.4 percent among the three major source areas, the northwest mountain-front boundary, northeast regional-underflow boundary, and water table boundary; the comparison provides a semi-quantitative validation of the modified MODPATH code (appendix A). Improved estimates of source area contributions are possible using a finer particle release density and adjusting FRAC to better represent streamflow-infiltration recharge; however, preliminary testing of alternative particle density and FRAC could only produce small improvements in the estimated source-area contributions.

The sensitivity of backward particle-tracking results to changes in FRAC, the user-defined stopping criteria for weak source cells, is shown in figure 14A. Two additional MODPATH simulations were run setting FRAC equal to 0.1 and 0.9; these simulations were compared to the base-case simulation with FRAC=0.5. Recall that FRAC equal to 0.1 indicates that only 10 percent of the total outflow from a cell needs to originate from an internal source for a particle to be stopped, therefore increasing FRAC results in fewer particles being terminated at river cells (streamflow-infiltration recharge boundaries). For example, the change in FRAC from 0.1 to 0.9 results in a 7.5-percent reduction in the source area contributions from streamflow-infiltration recharge. This reduction was accompanied by a 7.2-percent increase in source area contributions from the northwest mountain-front boundary, with the largest increase (6.0 percent) coming from the Little Lost River source area. Particle pathlines passing through a weak source cell typically reached the northwest mountain-front boundary; however, a small percentage of these particles were orphans, and none reached the northeast regional-underflow boundary, as indicated by the insensitivity of source area contributions along the northeast boundary to variations in FRAC.

Groundwater source-area contributions for each model layer were determined by summing the number of particles terminating at a designated source-area boundary and dividing these by the total number of particles released within the model layer (fig. 14B and appendix G). Particle-release locations (figs. 15A, 15C, 15E, 15G, 15I, and 15K) in each model layer are color coded to identify the primary source of groundwater at the particle's release location. The resulting distribution of color-coded particles depicts the simulated distribution of source-area contributions to groundwater flow in the aquifer. The single-particle-release scenario identifies a single contributing source for each 2 by 2 block of cells. Alternative simulation scenarios that incorporate multiple particle releases within a cell block will, in many cases, show several contributing source areas for the cell block, particularly for cell blocks near boundaries separating the simulated source areas depicted in figure 15.

In model layer 1, tributary valley underflows from the Big Lost River, Little Lost River, and Birch Creek represent

15.6, 18.3, and 2.6 percent of the inflow sources; underflow from the northeast boundary represents 44.4 percent; and streamflow-infiltration recharge represents 13.6 percent. Orphans account for 5.5 percent of the total number of particles released in model layer 1. In model layer 2, tributary valley underflow represents 40.3 percent of the inflow sources, inflow from the regional aquifer water represents 49.1 percent, and streamflow infiltration represents 9.8 percent. Orphans account for 0.8 percent of the total number of particles released in model layer 2. In model layer 6, the deepest layer, the source distribution changes to 2.3, 3.8, and 11.4 percent for tributary valley underflow from the Big Lost River, Little Lost River, and Birch Creek, respectively, to 82.4 percent for regional aquifer underflow along the northeast boundary, and to 0.1 percent streamflow-infiltration recharge. Interestingly, components of streamflow-infiltration recharge, primarily from the Birch Creek (612) and Big Lost River sinks and playas (608–610) stream-reach source boundaries persist to considerable depth (figs. 15G and 15I) in the aquifer even though the source of this recharge is across the upper water-table boundary at a considerable distance away from its downgradient occurrence in model layers 4 and 5. In summary, source area simulations indicate that the percentage of regional groundwater from the northeast boundary increases with depth, and the percentage of tributary valley underflow and streamflow-infiltration recharge decreases with depth.

Travel velocities were commensurate with the aquifer's hydraulic conductivity distribution and gradually decrease with depth (fig. 2; figs. 15B, 15D, 15F, 15H, 15J, and 15L). In the upper model layers, low-velocity tributary valley and regional aquifer underflow water that inflows from the northern boundaries travels in a southerly direction until coming into contact with the high-velocity regional water that inflows from the northeast boundary and travels in a southwesterly direction (figs. 15B and 15D). A sharp refraction of flowlines occurs at the intersection of low- and high-velocity flows resulting from a contrast in hydraulic conductivity of almost 2 orders of magnitude between hydrogeologic zones 1 (11,700 ft/d) and 11 (227 ft/d) and is consistent with advective transport of particles under steady-state flow conditions. This contrast diminishes in the lower model layers where velocity differences between tributary valley and regional water are relatively small (figs. 15F, 15H, 15J, and 15L). The somewhat chaotic pattern of mapped travel velocities along flow paths is accentuated by the two-dimensional (2-D) rendering of particles traveling through a 3-D layered model grid. For example, particles released in model layer 2 (fig. 15D) travel through layers 1–4; particles released in model layers 3–5 (figs. 15F, 15H, 15J) travel through layers 1–5; and particles released in model layer 6 (fig. 15L) primarily travel through the deeper model layers 4–6. The absence of depth information is less problematic for particles released in model layer 1 (fig. 15B) because these particles primarily travel through layer 1.

MODPATH layered-grid simulations

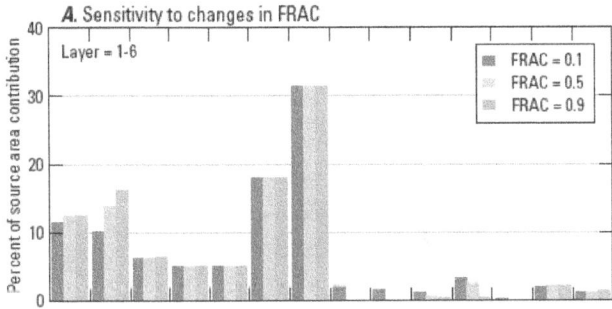

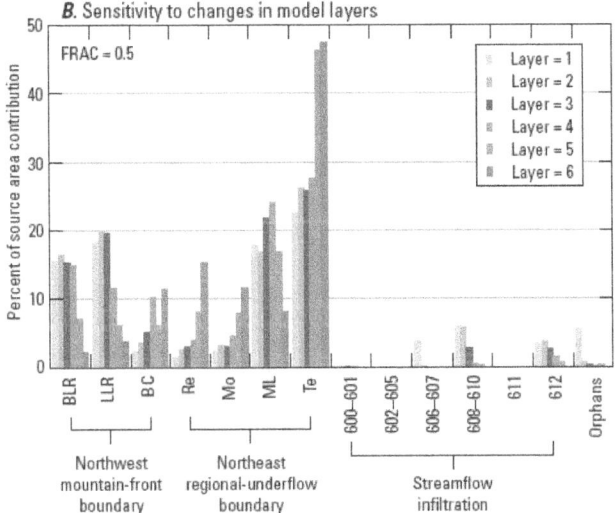

EXPLANATION

Source areas

BLR—Big Lost River valley
LLR—Little Lost River valley
BC—Birch Creek valley
Re—Reno section
Mo—Monteview section
ML—Mud Lake section
Te—Terreton section
600-601—Big Lost River stream reaches 600 and 601
602-605—Big Lost River spreading area, stream reaches 602 and 605
606-607—Big Lost River stream reaches 606 and 607
608-610—Big Lost River sinks and playas, stream reaches 608 and 610
611—Little Lost River stream reach 611
612—Birch Creek stream reach 612

Figure 14. Sensitivity of backward particle-tracking results to changes in (*A*) FRAC, the fraction of the total outflow from a cell contributed by an internal source and used to identify weak source cells; and (*B*) Layer, the model layer in which particles were released, Idaho National Laboratory and vicinity, Idaho. Results are shown as the percent source-area contributions of tributary valley, streamflow-infiltration recharge, and regional aquifer water for particles released in a layered-grid configuration within the model domain.

Comparison of Simulated Source Areas to Observed Source Areas

In model layers 1 and 2 the simulated position of the line separating groundwater derived from tributary valley underflow and streamflow-infiltration recharge sources (Type A water) from regional aquifer underflow sources (Type B water) crosses the 5 μg/L Li line[2] (note: the position of this boundary is defined for groundwater within the upper 200 ft of the aquifer) near the midpoint of its north-south trace (figs. 15*A* and 15*C*). North of this intersection the steady-state model overpredicts the eastern extent of tributary valley underflow from Birch Creek and streamflow-infiltration recharge from Birch Creek stream reach 612 and Big Lost River stream reach 610 (playa) (fig. 3). This overprediction is also indicated in model layer 3 at Site 14 (open to model layers 3 and 4) and in model layers 3 and 4 at well USGS 7 (open to model layers 3–6) (figs. 15*E*, 15*G*, and fig. 16; table 6). This overprediction occurs in hydrogeologic zones 11 and 44 in an area of the aquifer where simulated groundwater velocities are generally less than about 15 ft/d (figs. 15*B*, 15*D*, 15*F*, and 15*H*).

South of the intersection where the simulated position of the line separating Type A from Type B water in model layers 1 and 2 crosses the 5 μg /L Li line, the steady-state model underpredicts the eastern extent of tributary valley underflow from the Little Lost River and streamflow-infiltration recharge from Little Lost River stream reach 611 and Big Lost River stream reaches 602–605 (spreading areas), 606, 607, and 608–609 (playas). This underprediction occurs in an area of the aquifer where low-velocity groundwater (generally less than about 15 ft/d) in the sediment-rich rock of hydrogeologic zone 11 comes into contact with very high-velocity groundwater (60 to greater than 100 ft/d) in the sediment-poor rock of hydrogeologic zone 1 (model layers 1 and 2) and zone 2 (model layer 2).

In successively deeper model layers (figs. 15*E*, 15*G*, 15*I*, and 15*K*) the simulated position of the boundary separating tributary valley and streamflow-infiltration recharge sources from regional aquifer sources shifts progressively westward. Li concentration data from nine MLMS wells (figs. 7*A*, 7*B*, 16; table 6) indicate that the model overpredicts the western extent of groundwater derived from regional aquifer sources in model layers 3–5 at well USGS 103, model layers 1–5 at well USGS 105, model layers 1–5 at well USGS 108, model layer 5 at well USGS 132, model layers 3 and 4 at well USGS 133, and model layers 3 and 4 at well Middle 2050A. This overprediction is also indicated in model layers 3–5 at well Site 9 (open to model layers 3–5).

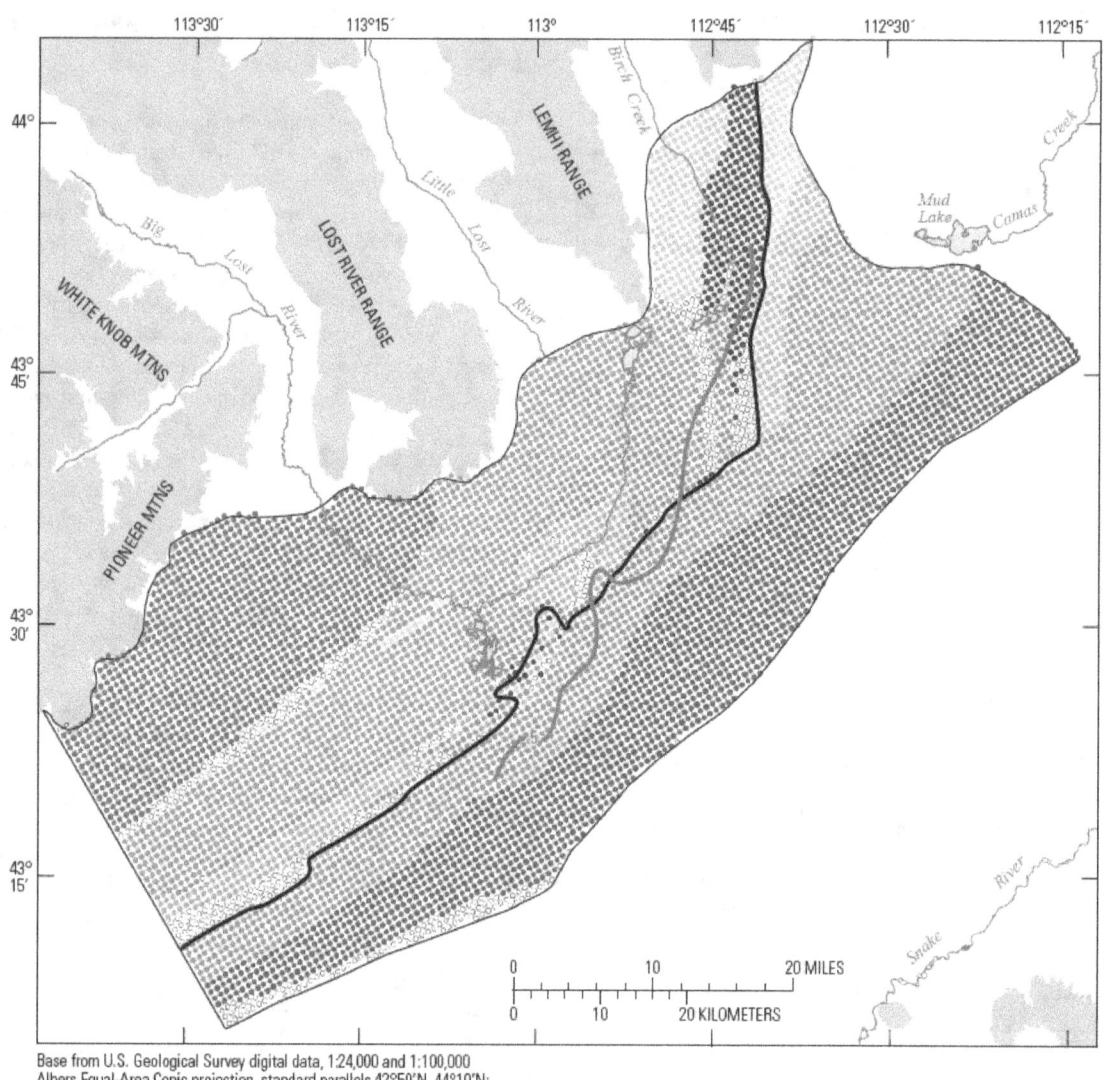

Base from U.S. Geological Survey digital data, 1:24,000 and 1:100,000
Albers Equal-Area Conic projection, standard parallels 42°50′N, 44°10′N;
central meridian 113°00′W; North American Datum of 1927.

A. Particles starting location and source area in model layer 1

EXPLANATION

Particles starting location and source area in model layer 1—layer is about 100 feet thick, varying with the water-table altitude. n is the number of particles that terminate in a designated source area.

Northwest mountain-front boundary source areas

● Big Lost River valley (n = 1,199)

● Little Lost River valley (n = 1,410)

● Birch Creek valley (n = 200)

Northeast regional-underflow boundary source areas

○ Northeast boundary Reno section (n = 125)

● Northeast boundary Monteview section (n = 188)

● Northeast boundary Mud Lake section (n = 1,370)

● Northeast boundary Terreton section (n = 1,734)

☐ Model area boundary

━ 5 ━ **5-microgram-per-liter lithium concentration isopleth**—Applies only to the upper 200 feet of the aquifer

Streamflow infiltration source areas

Big Lost River stream reaches 600 and 601 (n = 15)

● Big Lost River spreading area, stream reaches 602 and 605 (n = 0)

● Big Lost River stream reaches 606 and 607 (n = 302)

● Big Lost River sinks and playas, stream reaches 608 and 610 (n = 461)

● Little Lost River stream reach 611 (n = 0)

● Birch Creek stream reach 612 (n = 270)

Orphans

○ Particles that did not terminate in one of the specified source areas (n = 426)

━━━ **Major simulated interface between types A and B water**

Figure 15. Backward particle tracking starting locations, source areas, pathlines, and velocities for particles released in (*A, B*) model layer 1, (*C, D*) model layer 2, (*E, F*) model layer 3, (*G, H*) model layer 4, (*I, J*) model layer 5, and (*K. L*) model layer 6, Idaho National Laboratory and vicinity, Idaho. Particle pathlines are not shown for every release location in the simulation.

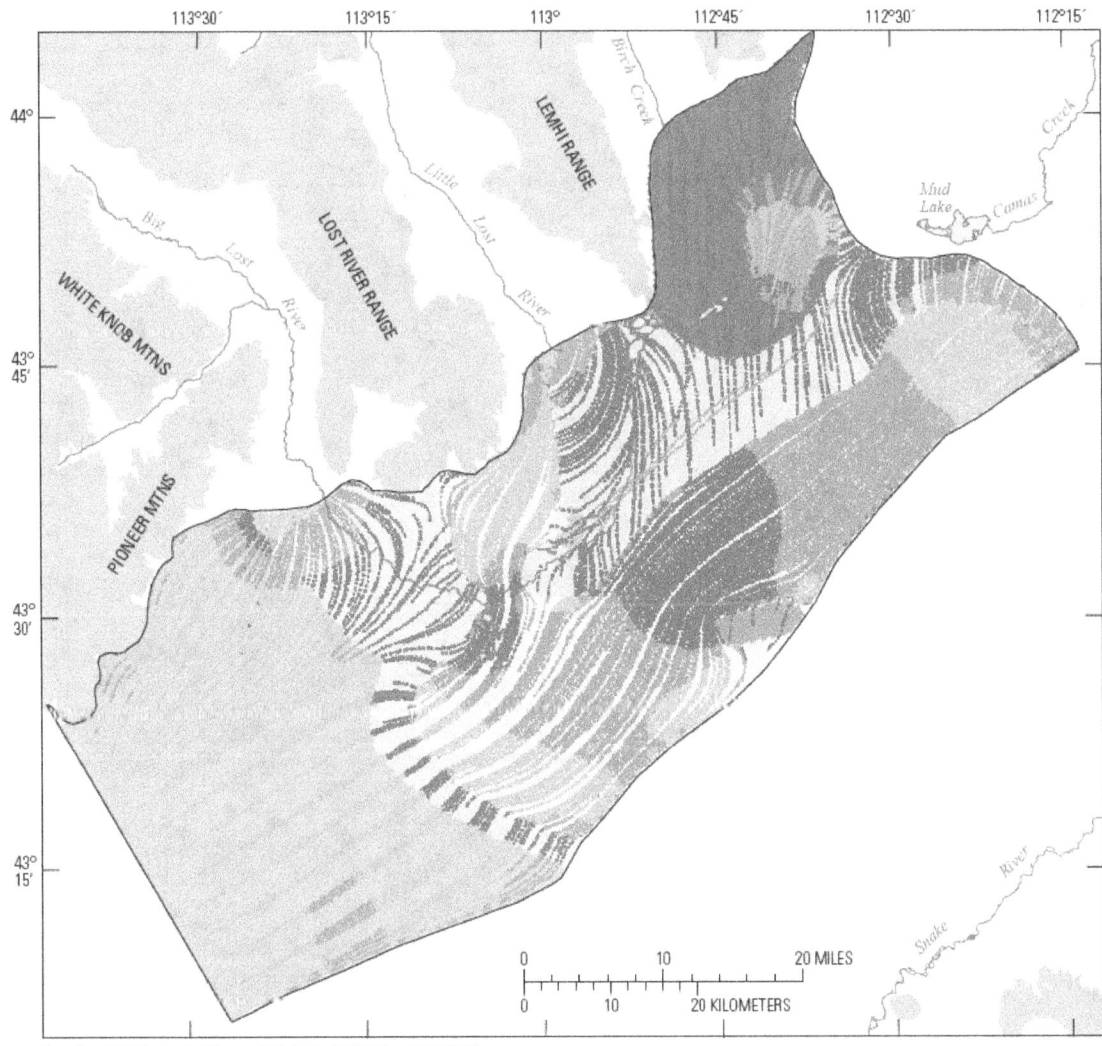

Base from U.S. Geological Survey digital data, 1:24,000 and 1:100,000
Albers Equal-Area Conic projection, standard parallels 42°50'N, 44°10'N;
central meridian 113°00'W; North American Datum of 1927.

B. Particle pathlines with starting locations in model layer 1

EXPLANATION

Particle pathlines with starting locations in model layer 1—particles
primarily traveling in model layer 1

····· Particles traveling at velocities greater than or equal to 0 and less than
or equal to 15 feet per day (ft/d)

Particles traveling at velocities greater than 15 and less than or equal
to 30 ft/d

Particles traveling at velocities greater than 30 and less than or equal
to 60 ft/d

····· Particles traveling at velocities greater than 60 and less than or equal
to 100 ft/d

····· Particles traveling at velocities greater than 100 ft/d

Model area boundary

Hydrogeologic zones, model layers 1

1—Younger rocks consisting of densely fractured basalt and interbedded sediment,
with a sediment thickness of generally less than 11 percent

2—Younger rocks consisting of massive, less densely fractured basalt and
interbedded sediment, with a sediment thickness of generally less than 11 percent

3—Intermediate-age rocks consisting of slightly altered fractured basalt and
sediment, with a sediment thickness of generally less than 11 percent

4—Intermediate-age rocks consisting of slightly altered fractured basalt and
sediment, with a sediment thickness of generally less than 11 percent

11—Younger rocks consisting of densely fractured basalt and interbedded sediment,
with a sediment thickness of generally more than 11 percent

22—Younger rocks consisting of massive, less densely fractured basalt and
interbedded sediment, with a sediment thickness of generally more than 11 percent

33—Intermediate-age rocks consisting of slightly altered fractured basalt and
sediment, with a sediment thickness of generally more than 11 percent

44—Intermediate-age rocks consisting of slightly altered fractured basalt and
interbedded sediment, with a sediment thickness of generally more than 11 percent

6—Silicic rocks, including rhyolite domes and andesite

Figure 15.—Continued

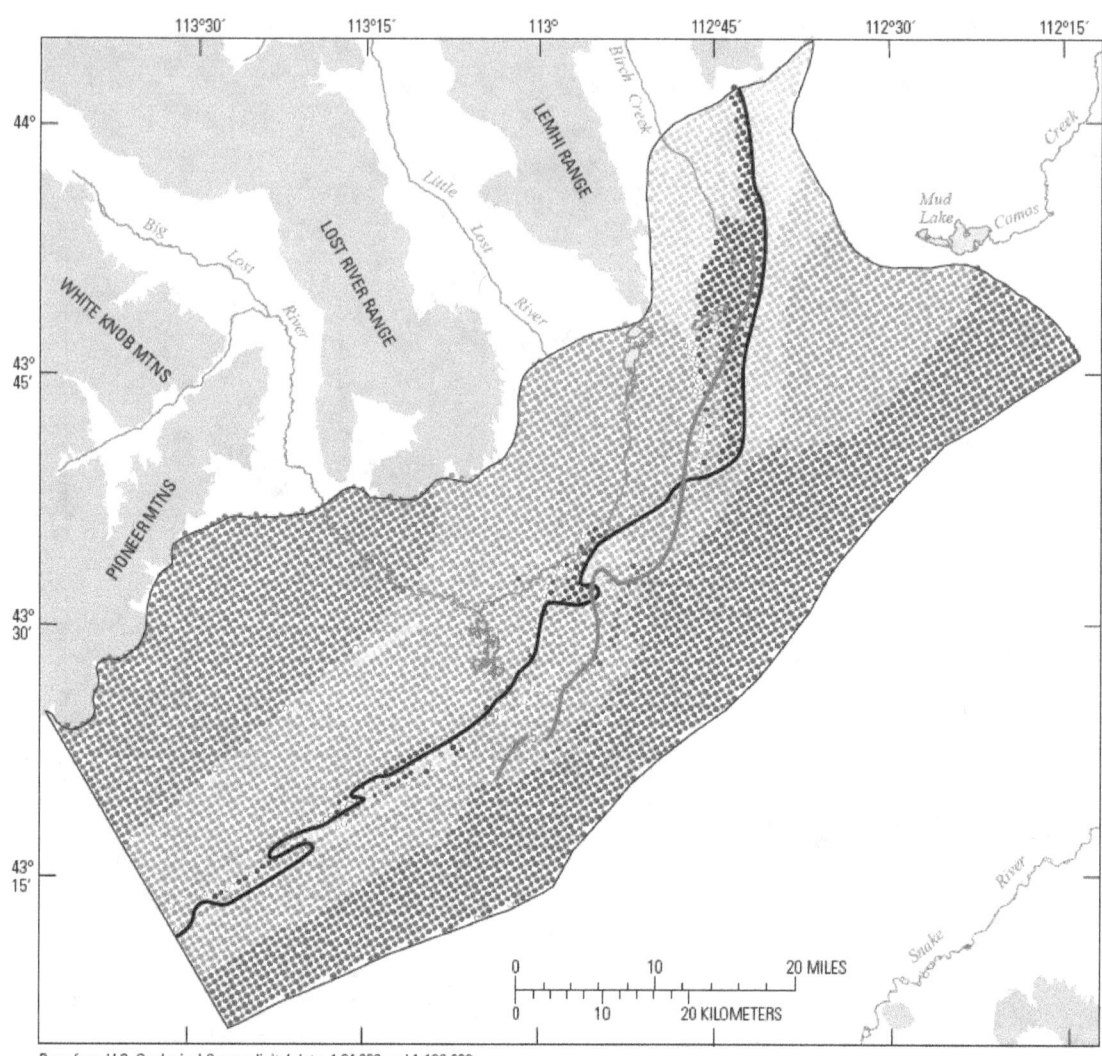

Base from U.S. Geological Survey digital data, 1:24,000 and 1:100,000
Albers Equal-Area Conic projection, standard parallels 42°50'N, 44°10'N;
central meridian 113°00'W; North American Datum of 1927.

C. Particles starting location and source area in model layer 2

EXPLANATION

Particles starting location and source area in model layer 2—layer is about 100 feet thick, varying with the water-table altitude.
n is the number of particles that terminate in a designated source area.

Northwest mountain-front boundary source areas

- Big Lost River valley (n = 1,272)
- Little Lost River valley (n = 1,543)
- Birch Creek valley (n = 288)

Northeast regional-underflow boundary source areas

- Northeast boundary Reno section (n = 205)
- Northeast boundary Monteview section (n = 254)
- Northeast boundary Mud Lake section (n = 1,306)
- Northeast boundary Terreton section (n = 2,018)

☐ Model area boundary

━ 5 ━ **5 microgram per liter lithium concentration
isopleth**—Applies only to the upper 200 feet of
the aquifer

Streamflow infiltration source areas

- Big Lost River stream reaches 600 and 601 (n = 13)
- Big Lost River spreading area, stream reaches 602 and 605 (n = 0)
- Big Lost River stream reaches 606 and 607 (n = 8)
- Big Lost River sinks and playas, stream reaches 608 and 610 (n = 442)
- Little Lost River stream reach 611 (n = 0)
- Birch Creek stream reach 612 (n = 294)

Orphans

○ Particles that did not terminate in one of the specified source areas (n = 65)

━━━ **Major simulated interface between types A
and B water**

Figure 15.—Continued

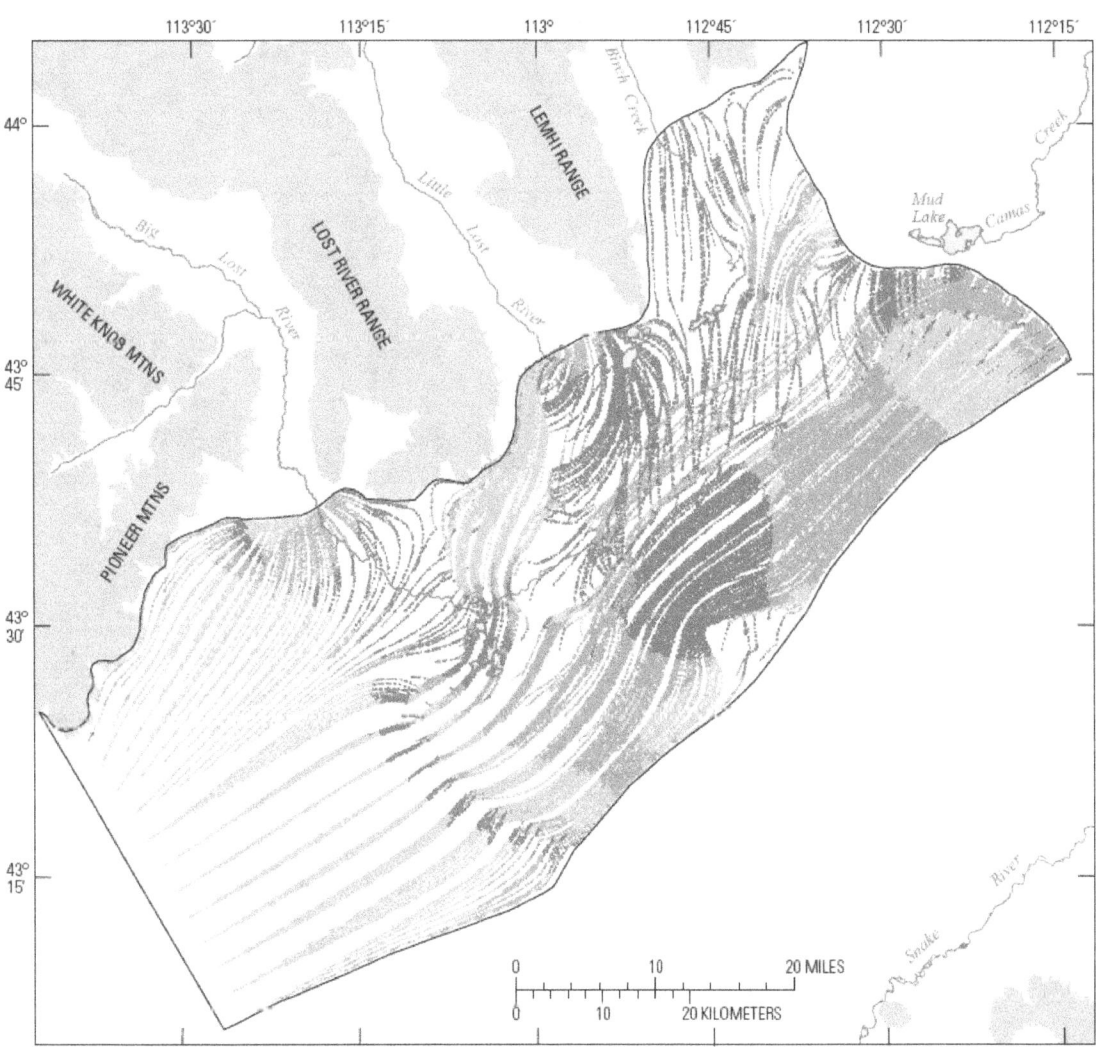

Base from U.S. Geological Survey digital data, 1:24,000 and 1:100,000
Albers Equal-Area Conic projection, standard parallels 42°50'N, 44°10'N;
central meridian 113°00'W; North American Datum of 1927.

D. Particle pathlines with starting locations in model layer 2

EXPLANATION

Particle pathlines with starting locations in model layer 2—particles
primarily traveling in model layer 1–4

.... Particles traveling at velocities greater than or equal to 0 and less than
or equal to 15 feet per day (ft/d)

.... Particles traveling at velocities greater than 15 and less than or equal
to 30 ft/d

.... Particles traveling at velocities greater than 30 and less than or equal
to 60 ft/d

.... Particles traveling at velocities greater than 60 and less than or equal
to 100 ft/d

.... Particles traveling at velocities greater than 100 ft/d

☐   Model area boundary

Figure 15.—Continued

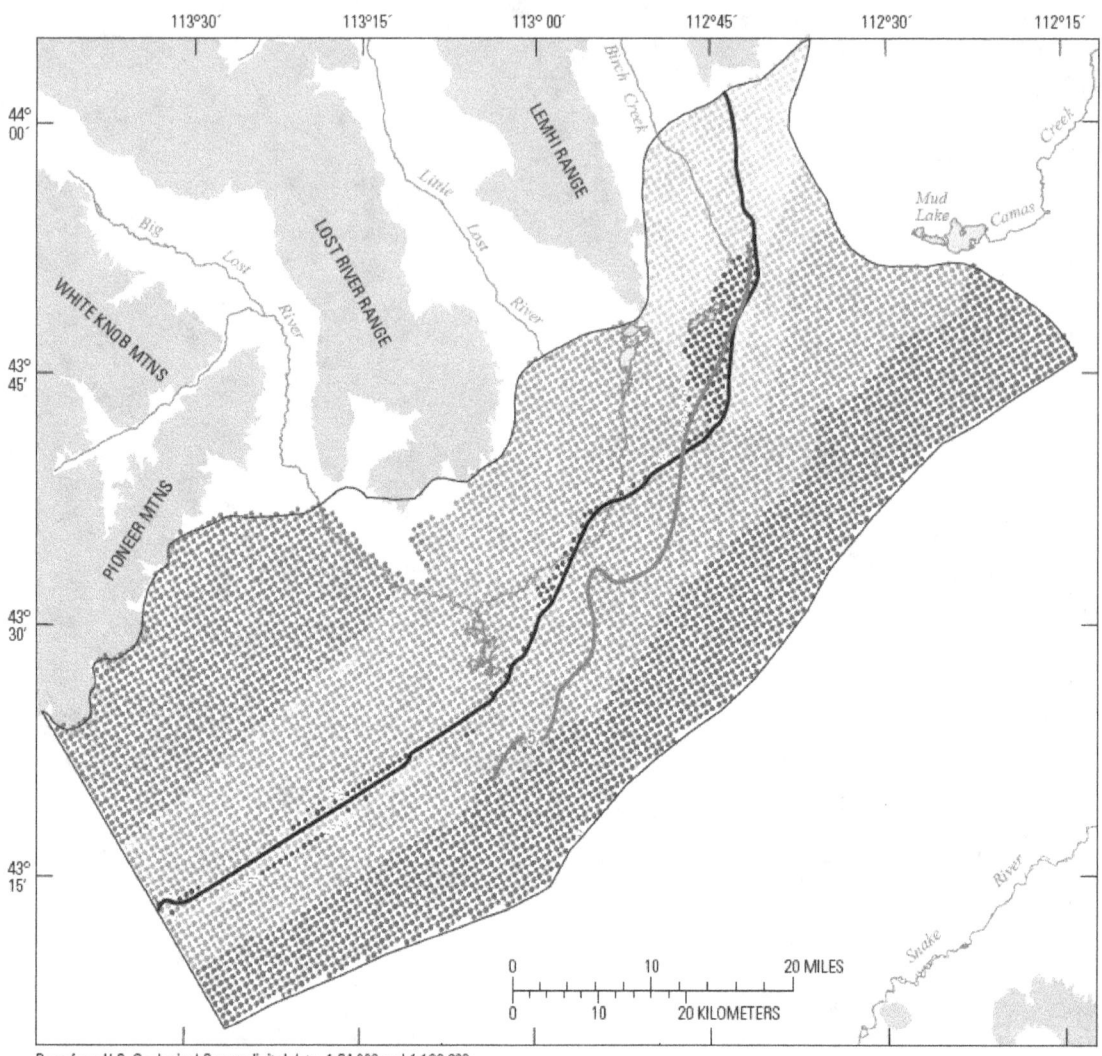

Base from U.S. Geological Survey digital data, 1:24,000 and 1:100,000
Albers Equal-Area Conic projection, standard parallels 42°50′N, 44°10′N;
central meridian 113°00′W; North American Datum of 1927.

E. Particles starting location and source area in model layer 3

EXPLANATION

Particles starting location and source area in model layer 3—layer is 0 to100 feet thick, about 200 to 300 feet below the water table. n is the number of particles that terminate in a designated source area.

Northwest mountain-front boundary source areas

● Big Lost River valley (n = 1,161)

● Little Lost River valley (n = 1,483)

● Birch Creek valley (n = 396)

Northeast regional-underflow boundary source areas

● Northeast boundary Reno section (n = 234)

● Northeast boundary Monteview section (n = 237)

● Northeast boundary Mud Lake section (n = 1,646)

● Northeast boundary Terreton section (n = 1,954)

◻ **Model area boundary**

━ 5 ━ **5 microgram per liter lithium concentration isopleth**—Applies only to the upper 200 feet of the aquifer

Streamflow infiltration source areas

● Big Lost River stream reaches 600 and 601 (n = 10)

● Big Lost River spreading area, stream reaches 602 and 605 (n = 0)

● Big Lost River stream reaches 606 and 607 (n = 2)

● Big Lost River sinks and playas, stream reaches 608 and 610 (n = 225)

● Little Lost River stream reach 611 (n = 0)

● Birch Creek stream reach 612 (n = 199)

Orphans

○ Particles that did not terminate in one of the specified source areas (n = 24)

━━━ **Major simulated interface between types A and B water**

Figure 15.—Continued

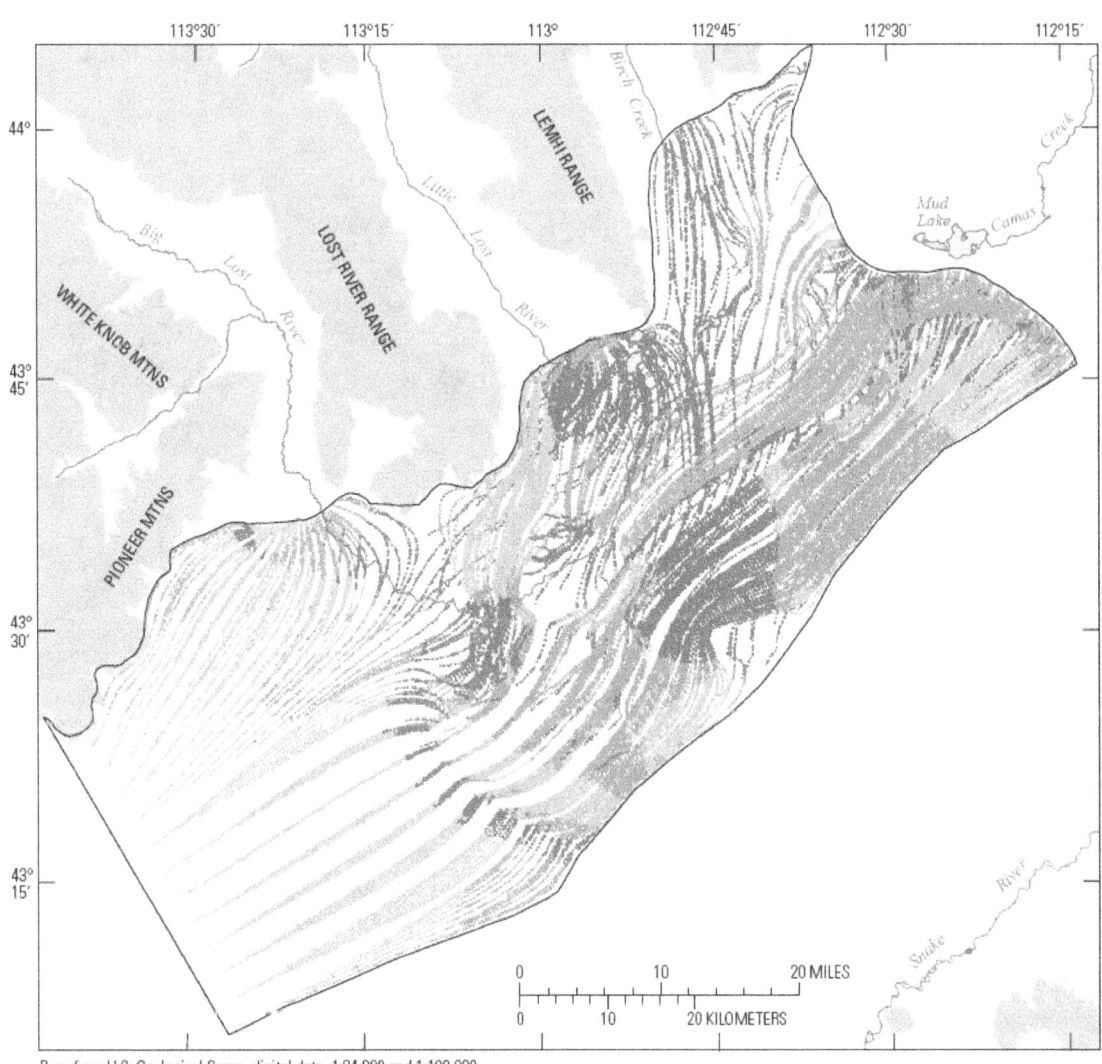

Base from U.S. Geological Survey digital data, 1:24,000 and 1:100,000
Albers Equal-Area Conic projection, standard parallels 42°50'N, 44°10'N;
central meridian 113°00'W; North American Datum of 1927.

F. Particle pathlines with starting locations in model layer 3

EXPLANATION

Particle pathlines with starting locations in model layer 3—particles primarily traveling in model layer 1–5

····· Particles traveling at velocities greater than or equal to 0 and less than or equal to 15 feet per day (ft/d)

····· Particles traveling at velocities greater than 15 and less than or equal to 30 ft/d

····· Particles traveling at velocities greater than 30 and less than or equal to 60 ft/d

····· Particles traveling at velocities greater than 60 and less than or equal to 100 ft/d

····· Particles traveling at velocities greater than 100 ft/d

☐ Model area boundary

Figure 15.—Continued

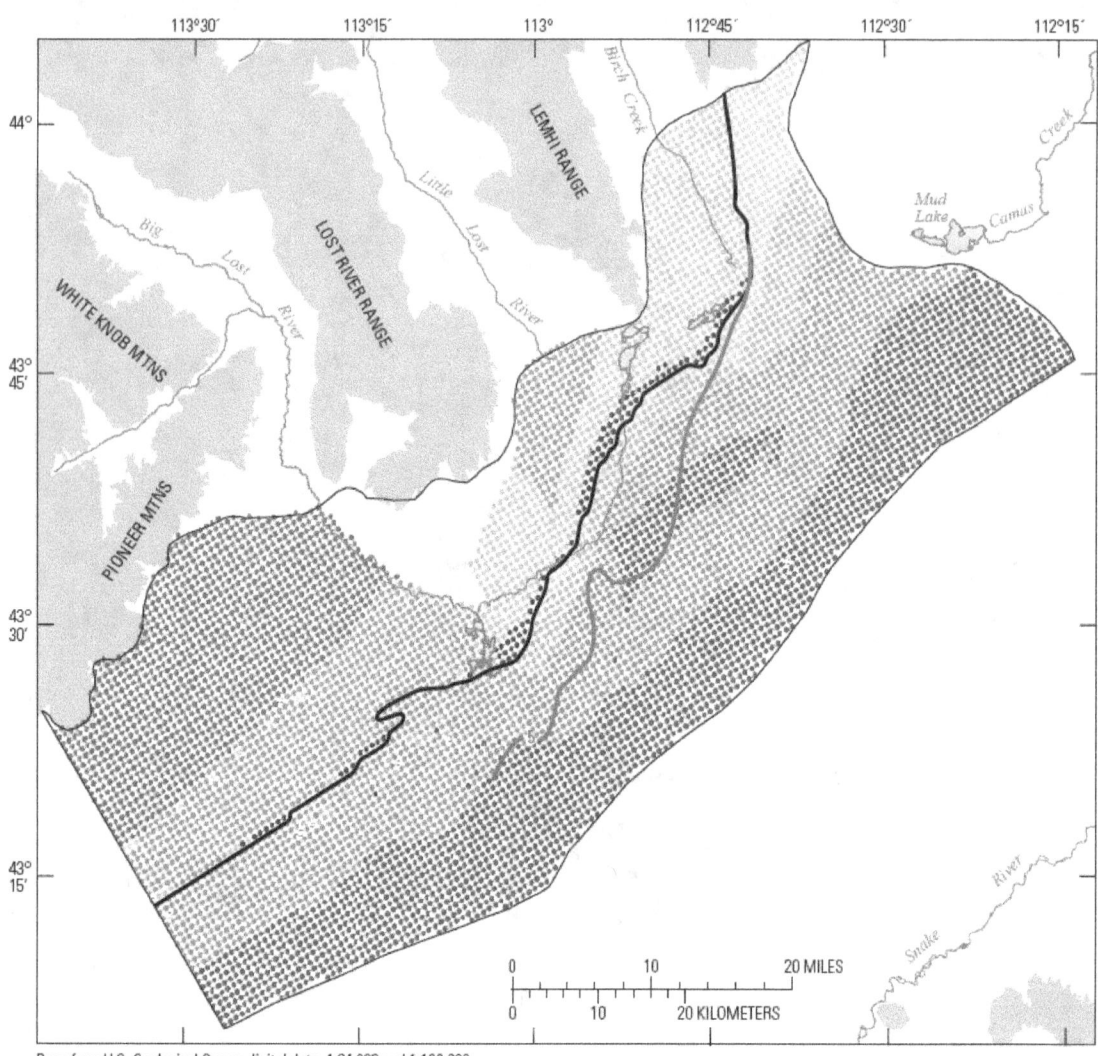

Base from U.S. Geological Survey digital data, 1:24,000 and 1:100,000
Albers Equal-Area Conic projection, standard parallels 42°50'N, 44°10'N;
central meridian 113°00'W; North American Datum of 1927.

G. Particles starting location and source area in model layer 4

EXPLANATION

Particles starting location and source area in model layer 4—layer is 0 to 200 feet thick, about 300 to 500 feet below the water
table. n is the number of particles that terminate in a designated source area.

Northwest mountain-front boundary source areas

- Big Lost River valley (n = 1,106)
- Little Lost River valley (n = 865)
- Birch Creek valley (n = 767)

Northeast regional-underflow boundary source areas

- Northeast boundary Reno section (n = 308)
- Northeast boundary Monteview section (n = 352)
- Northeast boundary Mud Lake section (n = 1,799)
- Northeast boundary Terreton section (n = 2,058

    ▭   **Model area boundary**

    ▬ 5 ▬   **5 microgram per liter lithium concentration
isopleth**—Applies only to the upper 200 feet of
the aquifer

Streamflow infiltration source areas

- Big Lost River stream reaches 600 and 601 (n = 9)
- Big Lost River spreading area, stream reaches 602 and 605 (n = 0)
- Big Lost River stream reaches 606 and 607 (n = 5)
- Big Lost River sinks and playas, stream reaches 608 and 610 (n = 48)
- Little Lost River stream reach 611 (n = 0)
- Birch Creek stream reach 612 (n = 107)

Orphans

- ○   Particles that did not terminate in one of the specified source areas (n = 8)

    ▬▬▬   **Major simulated interface between types A
and B water**

Figure 15.—Continued

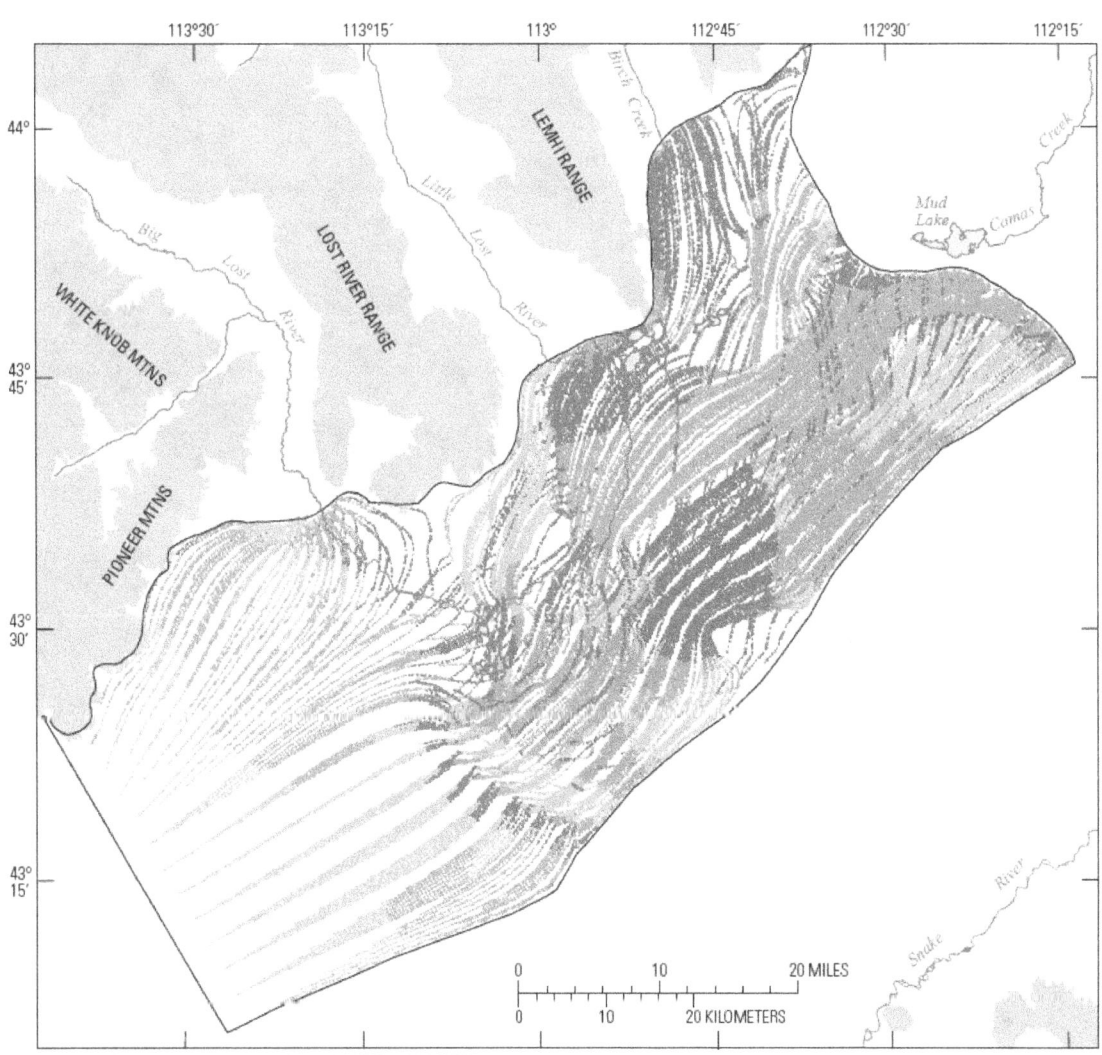

Base from U.S. Geological Survey digital data, 1:24,000 and 1:100,000
Albers Equal-Area Conic projection, standard parallels 42°50'N, 44°10'N;
central meridian 113°00'W; North American Datum of 1927.

H. Particle pathlines with starting locations in model layer 4

EXPLANATION

Particle pathlines with starting locations in model layer 4—particles
primarily traveling in model layer 1–5

· · · · ·   Particles traveling at velocities greater than or equal to 0 and less than
or equal to 15 feet per day (ft/d)

· · · · ·   Particles traveling at velocities greater than 15 and less than or equal
to 30 ft/d

· · · · ·   Particles traveling at velocities greater than 30 and less than or equal
to 60 ft/d

· · · · ·   Particles traveling at velocities greater than 60 and less than or equal
to 100 ft/d

· · · · ·   Particles traveling at velocities greater than 100 ft/d

[ ]   Model area boundary

Figure 15.—Continued

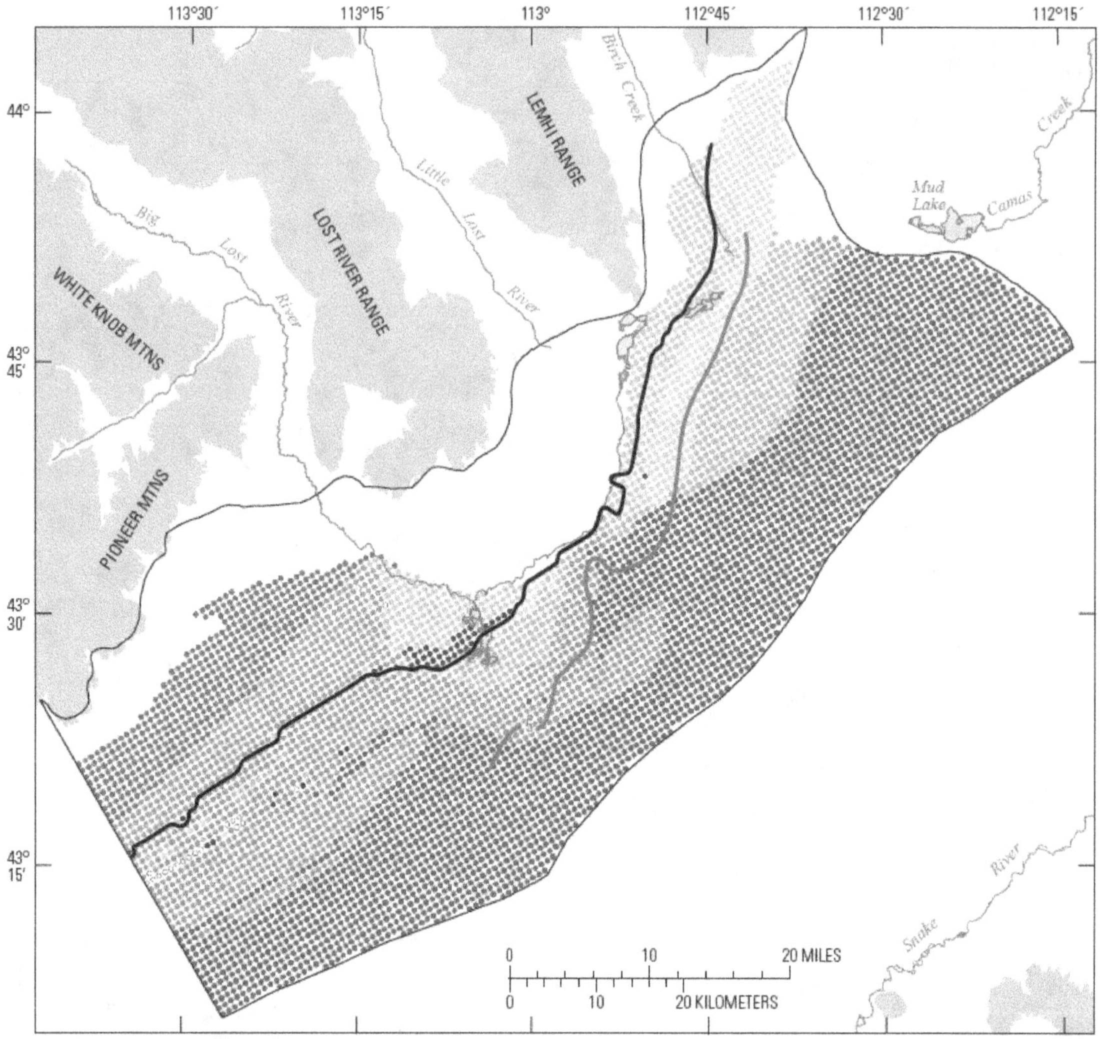

Base from U.S. Geological Survey digital data, 1:24,000 and 1:100,000
Albers Equal-Area Conic projection, standard parallels 42°50′N, 44°10′N;
central meridian 113°00′W; North American Datum of 1927.

I. Particles starting location and source area in model layer 5

EXPLANATION

Particles starting location and source area in model layer 5—layer is 0 to 300 feet thick, about 500 to 800 feet below the water table. n is the number of particles that terminate in a designated source area.

Northwest mountain-front boundary source areas

● Big Lost River valley (n = 438)

● Little Lost River valley (n = 374)

● Birch Creek valley (n = 373)

Northeast regional-underflow boundary source areas

● Northeast boundary Reno section (n = 494)

● Northeast boundary Monteview section (n = 485)

● Northeast boundary Mud Lake section (n = 1,028)

● Northeast boundary Terreton section (n = 2,825

☐ **Model area boundary**

═ 5 ═ **5 microgram per liter lithium concentration isopleth**—Applies only to the upper 200 feet of the aquifer

Streamflow infiltration source areas

Big Lost River stream reaches 600 and 601 (n = 0)

Big Lost River spreading area, stream reaches 602 and 605 (n = 0)

Big Lost River stream reaches 606 and 607 (n = 0)

● Big Lost River sinks and playas, stream reaches 608 and 610 (n = 18)

● Little Lost River stream reach 611 (n = 0)

● Birch Creek stream reach 612 (n = 49)

Orphans

○ Particles that did not terminate in one of the specified source areas (n = 19)

━━━ **Major simulated interface between types A and B water**

Figure 15.—Continued

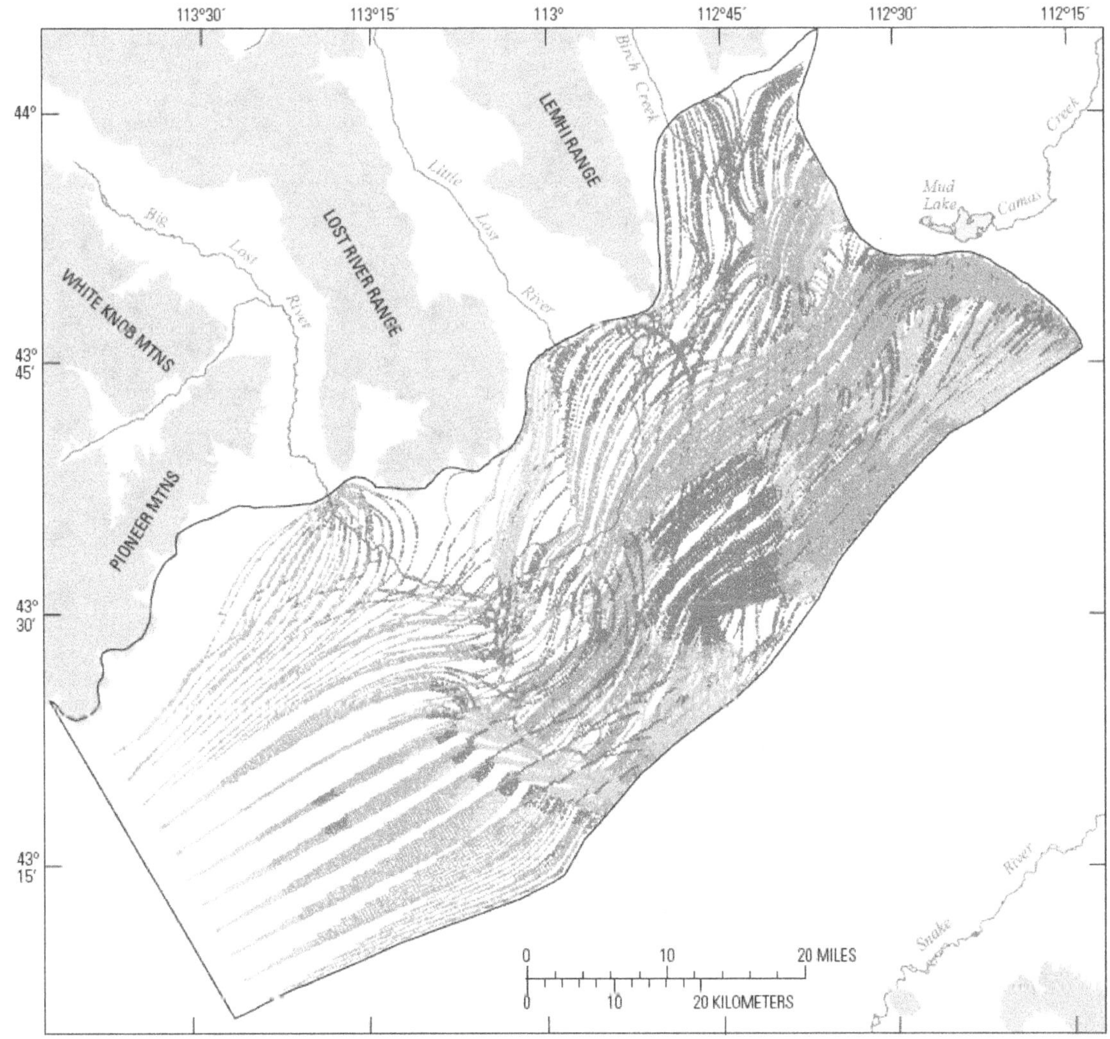

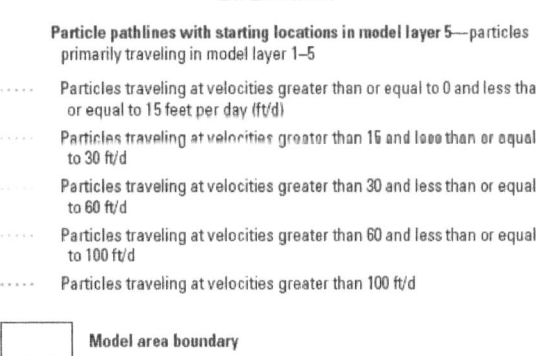

Base from U.S. Geological Survey digital data, 1:24,000 and 1:100,000
Albers Equal-Area Conic projection, standard parallels 42°50′N, 44°10′N;
central meridian 113°00′W; North American Datum of 1927.

J. Particle pathlines with starting locations in model layer 5

EXPLANATION

Particle pathlines with starting locations in model layer 5—particles
primarily traveling in model layer 1–5

⋯⋯ Particles traveling at velocities greater than or equal to 0 and less than
or equal to 15 feet per day (ft/d)

⋯⋯ Particles traveling at velocities greater than 15 and less than or equal
to 30 ft/d

⋯⋯ Particles traveling at velocities greater than 30 and less than or equal
to 60 ft/d

⋯⋯ Particles traveling at velocities greater than 60 and less than or equal
to 100 ft/d

⋯⋯ Particles traveling at velocities greater than 100 ft/d

☐ Model area boundary

Figure 15.—Continued

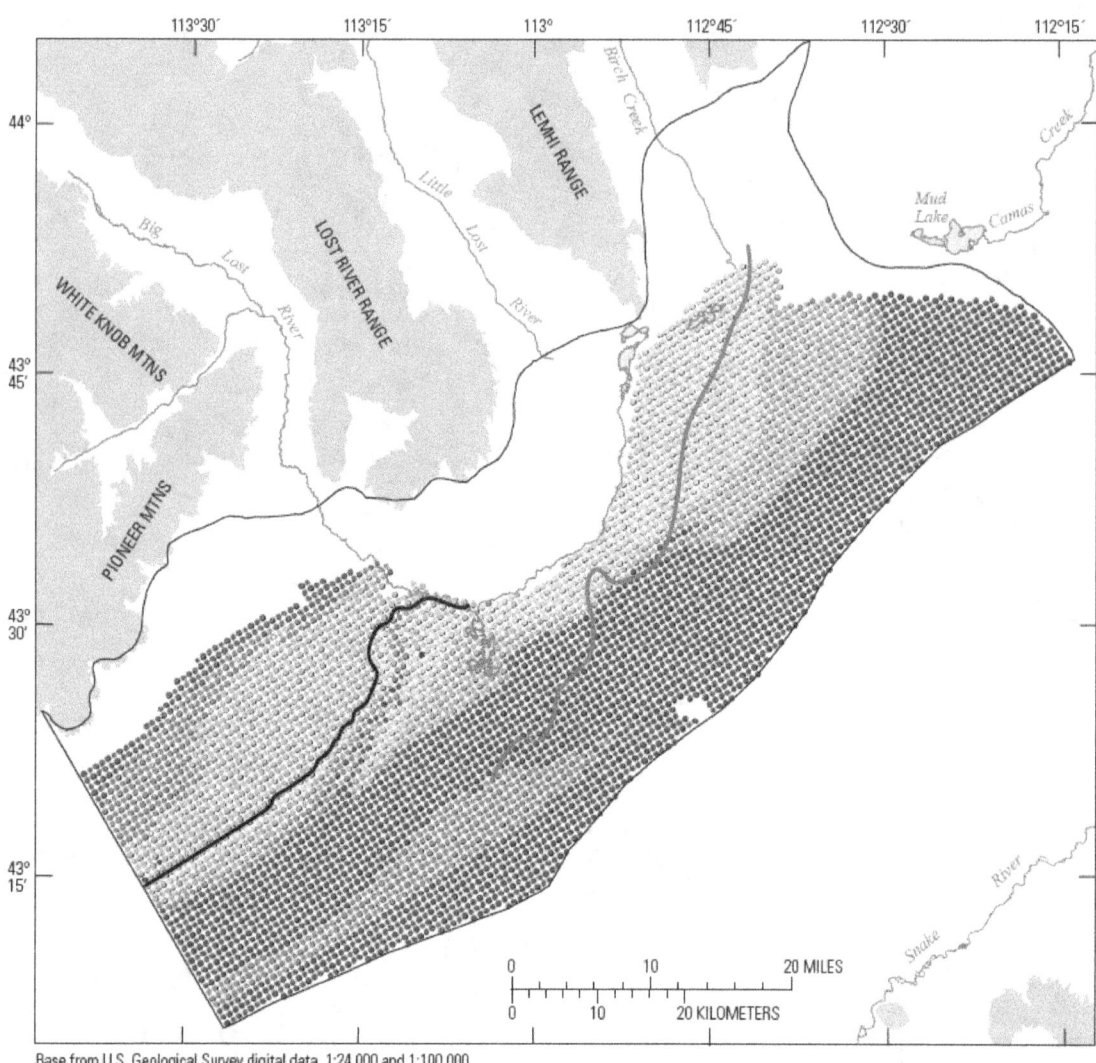

Base from U.S. Geological Survey digital data, 1:24,000 and 1:100,000
Albers Equal-Area Conic projection, standard parallels 42°50′N, 44°10′N;
central meridian 113°00′W; North American Datum of 1927.

K. Particles starting location and source area in model layer 6

EXPLANATION

Particles starting location and source area in model layer 6—layer is 0 to 3,229 feet thick, about 800 to 5,029 feet below the
water table. n is the number of particles that terminate in a designated source area.

Northwest mountain-front boundary source areas

- Big Lost River valley (n = 127)
- Little Lost River valley (n = 212)
- Birch Creek valley (n = 633)

Northeast regional-underflow boundary source areas

- Northeast boundary Reno section (n = 854)
- Northeast boundary Monteview section (n = 640)
- Northeast boundary Mud Lake section (n = 450)
- Northeast boundary Terreton section (n = 2,628

  ▢ **Model area boundary**

- — 5 — **5 microgram per liter lithium concentration
  isopleth**—Applies only to the upper 200 feet of
  the aquifer

Streamflow infiltration source areas

- Big Lost River stream reaches 600 and 601 (n = 0)
- Big Lost River spreading area, stream reaches 602 and 605 (n = 0)
- Big Lost River stream reaches 606 and 607 (n = 0)
- Big Lost River sinks and playas, stream reaches 608 and 610 (n = 0)
- Little Lost River stream reach 611 (n = 0)
- Birch Creek stream reach 612 (n = 4)

Orphans

- ○ Particles that did not terminate in one of the specified source areas (n = 2)

———— **Major simulated interface between types A
and B water**

Figure 15.—Continued

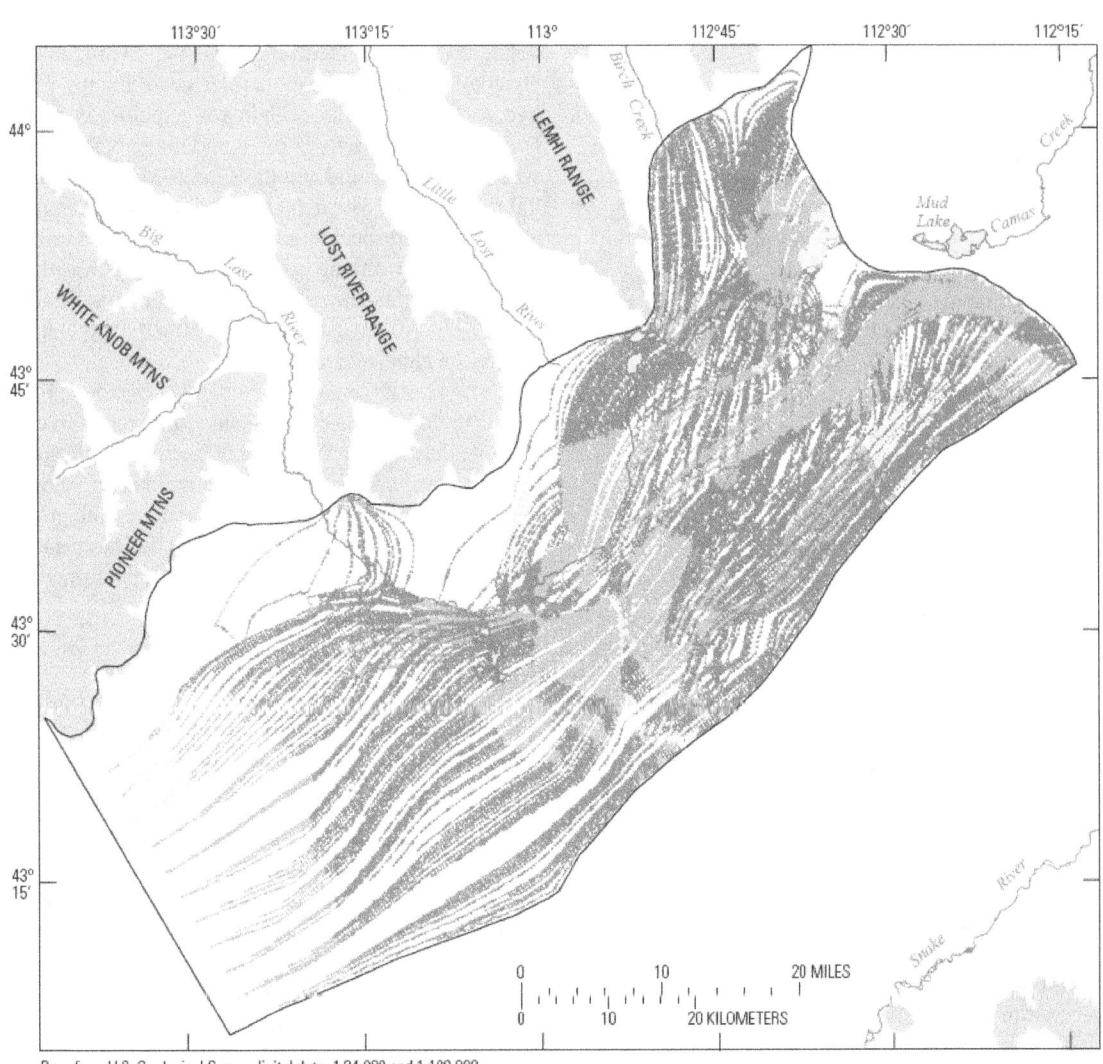

Base from U.S. Geological Survey digital data, 1:24,000 and 1:100,000
Albers Equal-Area Conic projection, standard parallels 42°50′N, 44°10′N;
central meridian 113°00′W; North American Datum of 1927.

L. Particle pathlines with starting locations in model layer 6

EXPLANATION

Particle pathlines with starting locations in model layer 6—particles
primarily traveling in model layer 4–6

..... Particles traveling at velocities greater than or equal to 0 and less than
or equal to 15 feet per day (ft/d)

Particles traveling at velocities greater than 15 and less than or equal
to 30 ft/d

..... Particles traveling at velocities greater than 30 and less than or equal
to 60 ft/d

..... Particles traveling at velocities greater than 60 and less than or equal
to 100 ft/d

..... Particles traveling at velocities greater than 100 ft/d

Model area boundary

Figure 15.—Continued

Evaluation of the Six Model Layer Simulation Results

Comparisons of simulated to observed source areas provide an independent, integrated way of evaluating model performance at a regional scale that is compatible with the model's simplified representation of the aquifer as a porous-media equivalent. The estimates of hydraulic conductivity and vertical anisotropy used in the source-tracking model were optimized using inverse modeling methods to minimize differences between simulated and observed heads (Ackerman and others, 2010). The relevance of the computational results from inverse modeling is dependent on (1) the spatial density of measurement wells, (2) the measurement error of observed heads, (3) a realistic representation of the hydrogeologic framework, (4) the reliability of inflow estimates across model boundaries, and (5) the validity of the steady-state assumption.

Source area comparisons indicate several shortcomings in the way the model represents flow in the aquifer. The eastward movement of tributary valley underflow and streamflow-infiltration recharge is overpredicted in the north-central part of the model area and underpredicted in the central part of the model area (fig. 16; table 6). Data are not available to evaluate source area simulations in the southwestern part of the model area.

For the most part, model inconsistencies can be attributed to large contrasts in hydraulic conductivity between hydrogeologic zones and the use of a single, model-wide value of vertical anisotropy. This evaluation approach assumes there are no gross misrepresentations of inflow across model boundaries and that the steady-state assumption is valid. A test of alternative estimates of flow to or from the aquifer indicated that only a 20-percent decrease in the largest flow, the northeast boundary underflow, resulted in a significant change to a calibrated parameter value, although major features of the flow system, such as the dominance of horizontal flow, were not affected (Ackerman and others, 2010, p. 132). More than 50 years of water-level observations in more than 100 wells (with variable lengths of record) indicate that the aquifer is never truly at steady state, but the available data indicate that the steady-state assumption is most closely approximated by water table conditions in 1980 (Ackerman and others, 2006, p. 36).

Because water flows through a porous medium along the path of least resistance, flowlines will use high-permeability formations as conduits and traverse low-permeability formations by the shortest path (Freeze and Cherry, 1979, p. 173). This behavior is reflected in the way the model simulates groundwater movement into, through, and across each hydrogeologic zone in response to differences in model-derived estimates of horizontal and vertical hydraulic conductivity. Simulated velocities in all hydrogeologic zones are commensurate with model-derived estimates of their hydraulic conductivity, with lower velocities in zones where hydraulic conductivity is small and higher velocities in zones where hydraulic conductivity is large. Velocities ranged from 0.0 to 626.1 ft/d with a median and interquartile range of 26.4 and 45.6 ft/d, respectively. The primary reason velocities varied so much among hydrogeologic zones was the 2.1 order-of-magnitude range of simulated hydraulic conductivity values. The 0.03 to 0.15 range for effective porosity also influenced simulated velocities.

The very large value of vertical anisotropy (14,800) results in flow that is predominantly horizontal in each hydrogeologic zone. Within each hydrogeologic zone horizontal volumetric fluxes and groundwater velocities are much larger than vertical volumetric fluxes and groundwater velocities. Volumetric flux, as defined by Darcy's law for a zonal homogeneous transversely isotropic formation in a Cartesian coordinate system, is defined as

$$q_x = -K_{h,i}\frac{\partial h}{\partial x}, \ q_y = -K_{h,i}\frac{\partial h}{\partial y}, \ q_z = -K_{z,i}\frac{\partial h}{\partial z} = -\frac{K_{h,i}}{a_i}\frac{\partial h}{\partial z} \quad (5)$$

where

q_x, q_y, q_z are values of volumetric flux along the x, y (horizontal directions), and z (vertical direction) coordinate axes [L/T];

$K_{h,i}, K_{z,i}$ are values of hydraulic conductivity in the horizontal and vertical directions of hydrogeologic zone i [L/T];

h is the hydraulic head [L]; and

a_i is the vertical anisotropy in hydrogeologic zone i [dimensionless].

The average fluid velocity within pores is then given as

$$v_x = \frac{q_x}{n_i}, \ v_y = \frac{q_y}{n_i}, \ v_z = \frac{q_z}{n_i} \quad (6)$$

where

v_x, v_y, v_z are values of average velocity along the x, y, and z coordinate axes [L/T]; and

n_i is the effective porosity of hydrogeologic zone i [L^3/L^3].

---

[2] The 5-microgram-per-liter lithium line defines the western boundary of the mixing zone in the upper 200 feet of the aquifer. Lithium concentrations for groundwater and thermal spring water in the study area indicate a sharp break in Type A water from Type B water on the western boundary of the mixing zone. The boundary on the eastern side of the mixing zone is not well defined.

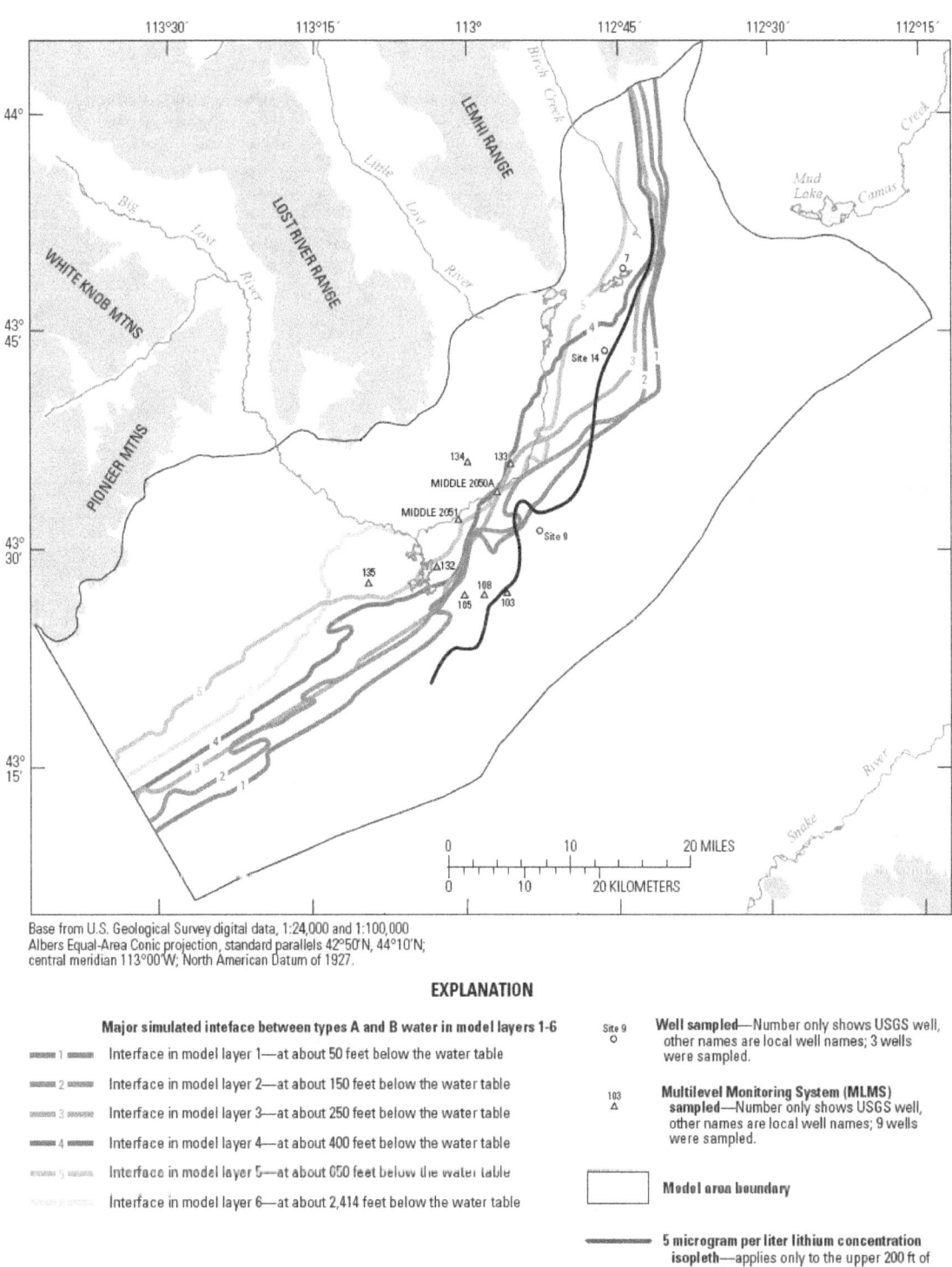

Base from U.S. Geological Survey digital data, 1:24,000 and 1:100,000
Albers Equal-Area Conic projection, standard parallels 42°50'N, 44°10'N;
central meridian 113°00'W; North American Datum of 1927.

EXPLANATION

Major simulated inteface between types A and B water in model layers 1-6

1 — Interface in model layer 1—at about 50 feet below the water table

2 — Interface in model layer 2—at about 150 feet below the water table

3 — Interface in model layer 3—at about 250 feet below the water table

4 — Interface in model layer 4—at about 400 feet below the water table

5 — Interface in model layer 5—at about 650 feet below the water table

6 — Interface in model layer 6—at about 2,414 feet below the water table

Site 9
○ **Well sampled**—Number only shows USGS well, other names are local well names; 3 wells were sampled.

103
△ **Multilevel Monitoring System (MLMS) sampled**—Number only shows USGS well, other names are local well names; 9 wells were sampled.

☐ **Model area boundary**

5 microgram per liter lithium concentration isopleth—applies only to the upper 200 ft of the aquifer

Figure 16. Simulated interface between groundwater derived primarily from tributary valley underflow and streamflow-infiltration recharge from groundwater derived primarily from regional aquifer underflow in model layers 1 through 6, Idaho National Laboratory and vicinity, Idaho.

Table 6. Summary of observed and simulated sources of groundwater in model layers 1 through 6, Idaho National Laboratory and vicinity, Idaho.

[**Layer depth:** The model layer depth in feet below the water table (ft bwt). **Obs:** The observed source of groundwater. **Sim:** The simulated source of groundwater. **A:** Groundwater derived from tributary valley underflow and streamflow-infiltration recharge. **B:** Groundwater derived from regional aquifer underflow. Bolded entry indicates simulated water source that is not consistent with observed water source. **Abbreviations:** –, not available; OMD, outside model domain]

Model layer	Layer depth (ft bwt)	USGS 7		USGS 103		USGS 105		USGS 108		USGS 132		USGS 133	
		Obs[1]	Sim	Obs	Sim	Obs	Sim	Obs	Sim	Obs	Sim	Obs	Sim
1	0–100	–	A	–[2]	B	A	**B**	A	**B**	A	A	A	A
2	100–200	–	A	B[2]	B	A	**B**	A	**B**	A	A	A	A
3	200–300	B	**A**	A	**B**	A	**B**	A	**B**	A	A	A	A,B
4	300–500	B	**A**	A	**B**	A	**B**	A	**B**	A	A	A	**B**
5	500–800	B	B	A	**B**	A	**B**	A	**B**	A	**B**	–	OMD
6	800–5,029	B	B	–	B	–	B	–	B	–	B	–	OMD

Model layer	Layer depth (ft bwt)	USGS 134		USGS 135		MIDDLE 2050A		MIDDLE 2051		Site 9		Site 14	
		Obs	Sim	Obs	Sim	Obs	Sim	Obs	Sim	Obs[1]	Sim	Obs[1]	Sim
1	0–100	A	A	A	A	A	A	A	A	–	B	–	A
2	100–200	A	A	A	A	A	A	A	A	–	B	–	A
3	200–300	A	A	A	A	A	**A,B**	A	A	A	**B**	B	**A**
4	300–500	A	A	A	A	A	**B**	A	A	A	**B**	B	B
5	500–800	–	OMD	–	A	A	OMD	A	OMD	A	**B**	–	B
6	800–5,029	–	OMD	–	B	A	OMD	A	OMD	–	B	–	B

[1] Open-hole completion with well casing extending down to depths greater than 200 feet below the water table.

[2] Water samples collected from well USGS 103 prior to Multilevel Monitoring System installation indicate groundwater that is a mixture of A and B in the upper 200 feet of the aquifer.

Because a single vertical anisotropy for all hydrogeologic zones (a_{1-6} = 14,800) is specified in the model and a single effective porosity was used for each zone, the horizontal to vertical ratio of volumetric fluxes and groundwater velocities is constant for all zones.

The effects of large contrasts in hydraulic conductivity on simulated source areas in the aquifer are summarized in the five "Case" (comparison) discussions below:

Case 1: $K_{h,1}$ (11,700 ft/d) much larger than $K_{h,11}$ (227 ft/d)—facilitates the eastward movement of groundwater from hydrogeologic zone 11 into hydrogeologic zone 1 along the eastern boundary separating sediment-poor from sediment-rich rocks. In the northern part of the model area, inflow of regional aquifer water across the Mud Lake (ML) sector of the northeast boundary enters zone 11 and is quickly diverted towards zone 1 and away from the area where the model overpredicts the eastern extent of tributary valley underflow and streamflow-infiltration recharge in model layers 1–4

(fig. 16). Farther south along this boundary, slow moving groundwater (less than 15 ft/d) in zone 11 comes into contact with very fast moving groundwater (from 30 to greater than 100 ft/d) in zone 1 (fig. 15) and the model underpredicts the eastern extent of groundwater derived from tributary valley underflow and streamflow-infiltration recharge. Flow (from hydrogeologic zone 11) across this boundary is sharply refracted in model layers 1 and 2, thus limiting the eastward movement of tributary valley underflow and streamflow-infiltration recharge in these two model layers. Zone 11 is absent below model layer 3 (fig. 2) and the underprediction of tributary valley underflow and streamflow-infiltration recharge in model layers 3, 4, and 5 in the north-central part of the model area can be attributed to downward flow of regional aquifer water from zone 11 into zone 22 (described in Case 3) and in the central part of the model areas to westward flow of regional aquifer water from zone 2 into zone 22 (described in Case 4).

In the western part of the model area, inflows entering zone 11 from the Little Lost River valley and stream reach 611 are quickly diverted (fig. 15) towards zone 1 and away from the area where the model underpredicts the eastern extent of tributary valley underflow and streamflow-infiltration recharge (fig. 16). Groundwater velocities in this area are greater than 30 ft/d. The model's representation of the hydraulic conductivity in this area may be too large. Hydraulic conductivity estimates from aquifer tests at well USGS 22 (0.009 ft/d) and well USGS 23 (300 ft/d) (Bartholomay and others, 1997) (figs. 1A and 1B) indicate that the hydraulic conductivity in the area east of the northwest mountain-front boundary and west of the boundary separating zone 1 from zone 11 may be much smaller than the model's best-fit estimate (11,400 ft/d). Groundwater temperatures measured in wells in this area are also high—19.8°C at well USGS 22, 15.4°C at well USGS 23, and 16.9°C at well USGS 19 (Busenberg and others, 2001, fig. 31). These temperatures are much higher than those in wells located within the sediment-rich rocks of zone 11 to the east (11.3 to 12.5°C) (Busenberg and others, 2001, fig. 31). The aquifer is thin at this location (figs. 2 and 5) and this could account for the higher temperatures; however, temperature profile data from well USGS 134 (June 2008) indicate that the geothermal gradient is linear, with temperatures ranging from 14.9°C at a depth of 368 ft below the water table to 12.8°C near the water table (Fisher and Twining, 2011, fig. 15 and table A1). The average temperature of groundwater at well USGS 134 is 13.7°C. The linear temperature gradient at well USGS 134 suggests that heat flow in the area east of the northwest mountain-front boundary and west of the boundary separating zone 11 from zone 1 is primarily conductive and that the convective component of heat flow, attributable to groundwater flow, is minimal. Collectively, aquifer-test estimates of hydraulic conductivity and the temperature data indicate that the hydraulic conductivity in this area likely is much smaller than the model-derived estimate (11,700 ft/d) and is comparable to that of either zone 11 (287 ft/d) or that of zone 2 (384 ft/d).

Case 2: $K_{h,4}$ (9,980 ft/d) much larger than $K_{h,44}$ (285 ft/d) —facilitates the movement of groundwater from hydrogeologic zone 44 into hydrogeologic zone 4 and diverts the inflow of regional aquifer water across the Reno (Re) and Monteview (Mo) sections of the northeast boundary towards zone 1 and away from the area where the model overpredicts the eastern extent of tributary valley underflow and streamflow-infiltration recharge at well USGS 7 (open to model layers 3–6) in model layers 3 and 4 (table 6). Sampling in this well, whether bailed, thief sampled, or pumped has always produced a Type B water (appendix D). No water samples are available from model layers 1 and 2 in this well, but it is likely that water in these upper layers also is Type B water. The model adequately predicts the presence of Type B water in model layer 5 at well USGS 7.

Case 3: $K_{h,22}$ (4,780 ft/d) much larger than $K_{h,11}$ (227 ft/d)—facilitates downward flow from hydrogeologic zone 11 into hydrogeologic zone 22 because the vertical conductivity of zone 22 (0.32 ft/d) is much larger than the vertical conductivity of zone 11 (0.015 ft/d). Although this downward flow represents only a small fraction of the total flow in zone 22, its effect on simulated source areas in model layers 3 and 4 near well Site 14 (open to model layers 3 and 4) (table 6) can explain some of the model's overprediction of tributary valley underflow and streamflow-infiltration recharge in model layer 3 in the north-central part of the model area.

Case 4: $K_{h,22}$ (4,780 ft/d) much larger than $K_{h,2}$ (384 ft/d) —facilitates the westward movement of water from hydrogeologic zone 2 into hydrogeologic zone 22 and likely explains much of the overprediction of regional aquifer water in the central part of the model area and primarily within model layers 2–4 (fig. 16). As noted in the introduction, a $K_{h,22}$ that is much larger than $K_{h,2}$ is inconsistent with the original conceptualization of how sediment should affect this parameter and was an outcome of the model calibration process that could not be explained readily (Ackerman and others, 2010, p. 58). The effect of $K_{h,22}$ that is much larger than $K_{h,2}$ can be seen in the backward particle-tracking pathlines for particles released in model layers 2, 3, 4, 5, and 6 (figs. 15D, 15F, 15H, 15J, and 15L). Backward particle-tracking pathlines indicate that a major source of groundwater in the central part of the model area (figs. 15C, 15E, 15G, 15I, and 15K) is regional aquifer water that enters the aquifer along the Mud Lake (ML) and Terreton (Te) sections of the northeast boundary and diverts quickly from zone 2 into zone 22; the pathlines also indicate that groundwater moves from zone 2 into zone 22 along the boundary separating sediment-poor from sediment-rich rocks. The contributions of regional aquifer water increase in successively deeper model layers, which explain the westward shift of the boundary separating Type A from Type B water, and the overprediction of contributions from the regional aquifer in successively deeper model layers in the central part of the model layer.

Case 5: $K_{h,1}$ (11,700 ft/d) much larger than $K_{h,2}$ (384 ft/d) — inhibits both lateral and downward flow from hydrogeologic zone 1 into hydrogeologic zone 2 because the horizontal (11,700 ft/d) and vertical (0.79 ft/d) conductivities of zone 1 are much larger than the horizontal (384 ft/d) and vertical conductivities (0.03 ft/d) of zone 2. These relations account for the very high velocities (from 30 to greater than 100 ft/d) east of the boundary separating sediment-poor rock from sediment-rich rock and west of the southeast-flowline boundary (figs. 15B and 15D). The highest velocities (greater than 100 ft/d) occur in model layers 1 and 2 where regional aquifer underflow across the northeast boundary into zone 11 and subsequently into zone 1 is forced to flow upward and over zone 2 where it subcrops across the tops of model layers 2 and 3 (fig. 2).

An Evaluation of Mixed Water Types at Well NPR-W01

In this backward particle-tracking simulation, particles were released in model layers 1 and 2 at the location of well NPR-W01 in the south-central part of the INL (fig. 17). This simulation was used to identify and quantify the different sources of groundwater within the upper 200 ft of the aquifer near the 5 µg/L Li boundary that is interpreted to mark the separation of groundwater derived primarily from tributary valley underflow and streamflow-infiltration recharge from groundwater derived primarily from regional aquifer underflow.

In an earlier study, Schramke and others (1996) used the geochemical mass balance and mixing models NETPATH (Plummer and others, 1991) and PHREEQE (Parkhurst and others, 1980) to determine the sources of groundwater at well NPR-W01. These models included major and minor cations and anions, dissolved organic carbon (DOC) and dissolved inorganic carbon (DIC), and stable and radioactive isotopes (^{3}H, ^{14}C, δ^{13}C, δ^{34}S, δ^{18}O, and δD).

The chemical evolution of groundwater at well NPR-W01 was established on the basis of the chemistry of groundwater at wells along two assumed flow paths (fig. 17), one originating from the Little Lost River valley (flow path 1) and the other originating from the vicinity of the Birch Creek valley that included regional aquifer underflow (flow path 2). These flow paths were defined by the shape of the water table upgradient of well NPR-W01. Wells used in Schramke's analysis were selected based on trends in the chemical and isotopic composition of water and the proximity of the well location with respect to the trace of the flow paths. Nine wells were sampled in this study, four along flow path 1 and four along flow path 2, in addition to well NPR-W01. Wells referenced in this summary are those that produced the most consistent model to reproduce the chemistry of groundwater at well NPR-W01. Mixing-model results indicated that groundwater below the upper 10 ft of the aquifer at well NPR-W01 represents a mixture of water with about equal amounts of groundwater from the vicinities of well USGS 12 (48.1 to 48.8 percent) and well USGS 17 (47.2 to 49.2 percent) and streamflow infiltration from the Big Lost River channel (2.0 to 4.7 percent) presumably near well NPR-W01 (Schramke and others, 1996, tables 5 and 7). Both wells (USGS 12 and USGS 17) were shown to include water from Big Lost River streamflow-infiltration recharge as well as water from other sources, as described below.

NPR-W01 (a 520-ft-deep well with open intervals from 500 to 535 ft bls; water level 461 ft bls; model layer 1) and other wells used in this study were sampled in September and October 1990. Many of the wells used in this study are in an area of the aquifer where the transient effects of streamflow-infiltration recharge on water levels in the aquifer are large, and transient response times are short (Ackerman and others, 2010, figs. 10*F*, 11, and 12). This is particularly true for all wells along flow path 1 and wells near the lower end of flow path 2 (USGS 17 and NPR-W01). Prior to sampling in 1990, water levels at well USGS 12 had risen by about 15 ft as a result of streamflow in the Big Lost River from 1982 through 1986. From 1987 to 1990 there was no flow in the Big Lost River and water levels at well USGS 12 had declined by about 12 ft at the time that the Schramke study was conducted. Transient model simulations, using 1980 steady-state water levels to represent initial conditions, were able to reproduce the transient effects of streamflow-infiltration recharge from flow in the Big Lost River from 1981 through 1984. The largest water level rises in the transient simulation were centered beneath the Big Lost River sinks and playas (Ackerman and others, 2010, fig. 36).

The chemistry of groundwater at well USGS 12 (a 692-ft-deep well with open intervals from 587 to 692 ft bls; water level 326 ft bls; model layers 3 and 4) (flow path 1) could not be traced to a precise source using a mass balance and mixing model approach because of uncertain effects of contamination from upgradient agricultural sources in the Little Lost River valley (Schramke and others, 1996, p. 534). A geochemical mass balance study of the Little Lost River basin by Swanson and others (2002) documented the effects of agricultural contamination in wells sampled near the lower part of the basin. Busenberg and others (2001, p. 86) found well USGS 12 to consist mostly of local recharge with little, if any, regional aquifer water. Given the location of well USGS 12, its depth, well completion design, and time of sampling, it is reasonable to assume that the chemistry of water from this well probably represents a combination of Big Lost River infiltration recharge and alluvial-aquifer underflow from the Little Lost River valley.

The chemistry of groundwater at well USGS 17 (a 498-ft-deep well with open intervals from 438 to 445 ft bls and 496 to 498 ft bls; water level 353 ft bls; model layers 1 and 2) (flow path 2) was attributed to three sources, each contributing nearly equal amounts of water: (1) streamflow infiltration from the Big Lost River (35 percent), (2) alluvial aquifer-underflow from the Little Lost River valley (30 percent), and (3) groundwater inflow from the vicinity of well Site 14 (35 percent) (Schramke and others, 1996, tables 6 and 7).

Water at well Site 14 (a 717-ft-deep well with open intervals 535 to 716 ft bls; water level 269 ft bls; model layers 3 and 4) was interpreted to be a mixture of alluvial-aquifer underflow from Birch Creek valley and regional aquifer underflow to the northeast, although the Birch Creek valley component could not be established (Schramke and others, 1996, p. 536). This interpretation is not entirely consistent with the results of the mixing zone analysis discussed earlier in this report. The chemistry of water from well Site 14 (appendix D) indicates that it is Type B water. Its Li concentration is 11.5 µg/L and its B/Li concentration ratio is 3 (table 4). The mixing zone analysis suggests that the water from well Site 14 is regional groundwater that has been mixed (diluted) with irrigation return flow and streamflow-infiltration recharge in the Mud Lake and Terreton areas along the northeast model boundary and not with water from tributary valley underflow along the northwest mountain-front boundary.

In summary, the mixing models of Schramke indicate that groundwater at well NPR-W01 represents a mixture of water that can be sourced to alluvial-aquifer underflow from the Little Lost River valley (greater than about 14 percent), streamflow-infiltration recharge from the Big Lost River (less than about 69 percent), and inflow from the northeast boundary of about 17 percent that may contain a small percentage of alluvial-aquifer underflow from the Birch Creek valley.

In order to compare how particles in the model moved in comparison to relative percentages of source water chemistry from Schramke and others (1996), particles were released in the vicinity of well NPR-W01 (fig. 17). In this simulation, particle release locations were uniformly distributed within a circular area with a 2,000-ft radius with its center located at the NRP-W01 well; the particle density within this area is one particle for every 7×10^{-4} mi^2. A total of 9,028 particles were released from model layers 1 and 2 at random depths between 0 and 200 ft below the water table boundary. Release locations were extended horizontally beyond the 0.5-ft well radius and vertically beyond the well's open interval, from about the water table to 47 ft below the water table boundary, because the modeled spatial discretization is too coarse to precisely represent the geometry of this well.

Particle tracking shows that (1) tributary valley underflow is exclusively from the Little Lost River valley and represents 3.6 percent of all source area contributions, (2) underflow from the northeast boundary represents 35.3 percent, and (3) streamflow-infiltration recharge represents 49.4 percent. Source area contributions from the Big Lost River (stream reaches 606–607 and 608–610) make up most of the surface-water recharge at 48.8 percent, with the remaining 0.6 percent contributed from Birch Creek (stream reach 612). Orphaned particles account for the residual 11.7 percent of all source area contributions (fig. 17 and table 7).

The sensitivity of backward particle-tracking results to changes in FRAC is shown in figure 18A. Two additional MODPATH runs were made setting FRAC equal to 0.1 and 0.9; these runs were compared to the base-case run with FRAC equal to 0.5. A decrease in FRAC from 0.5 to 0.1 results in a 20.1 percent increase in the source area contribution from stream reaches 606–607, and a 13.4 percent decrease from stream reaches 608–610; generally, particle paths terminated further upstream in the Big Lost River with very few particles reaching Birch Creek. The distribution of source area contributions changed dramatically for FRAC equal to 0.9 with streamflow-infiltration recharge reduced to 11.2 percent, and a 33.7 percent increase in Little Lost River valley underflow. As expected, source area contributions from the northeast regional-underflow boundary were insensitive to changes in FRAC (table 7).

Particles released in the upper 50 ft of model layer 1 were primarily tracked to streamflow-infiltration recharge source areas (58.4 percent), with 41.6 percent orphaned. Source area contributions for particle releases in the lower half of model layer 1 (50–100 ft below the simulated water table) were almost exclusively tracked to streamflow-infiltration source areas (99.9 percent), with the remainder orphaned (0.1 percent). In model layer 2, particles released in the upper half of the layer (100–150 ft depth interval) resulted in contributions of 14.1 percent from the Little Lost River valley, 41.4 percent from the northeast underflow boundary, and 39.6 percent from streamflow infiltration, with orphans accounting for the remaining 11.7 percent. All particles released in the lower half of model layer 2 (150–200 ft depth interval) were from the northeast underflow boundary. In summary, water in the vicinity of well NPR-W01 is primarily composed of surface water from the Big Lost River in the upper part of the aquifer; the water transitions to regional water in the lower part of the aquifer. However, water in the upper part of the aquifer could also represent a mixture of water that is sourced from both streamflow-infiltration recharge and underflow from the Little Lost River valley; the mixing ratio is dependent on FRAC. Particle tracking results based on a larger FRAC value indicate an increase in the source area contribution from the Little Lost River valley (33.7 percent at FRAC=0.9).

The sensitivity of backward particle-tracking results to changes in the areal radius of the cylindrical volume defining particle release locations is shown in figure 18C. Two additional MODPATH runs were made setting the radius equal to 5,000 and 660 ft and compared to the base-case run (radius=2,000 ft). The total number of released particles was identical for each MODPATH run (n=9,028), resulting in areal particle densities of 0.02 and 8×10^{-5} mi^2 for radii of 5,000 and 660 ft, respectively. In general, the sensitivity of the source area contributions changed very little for large differences in radius (fig. 18C; table 7).

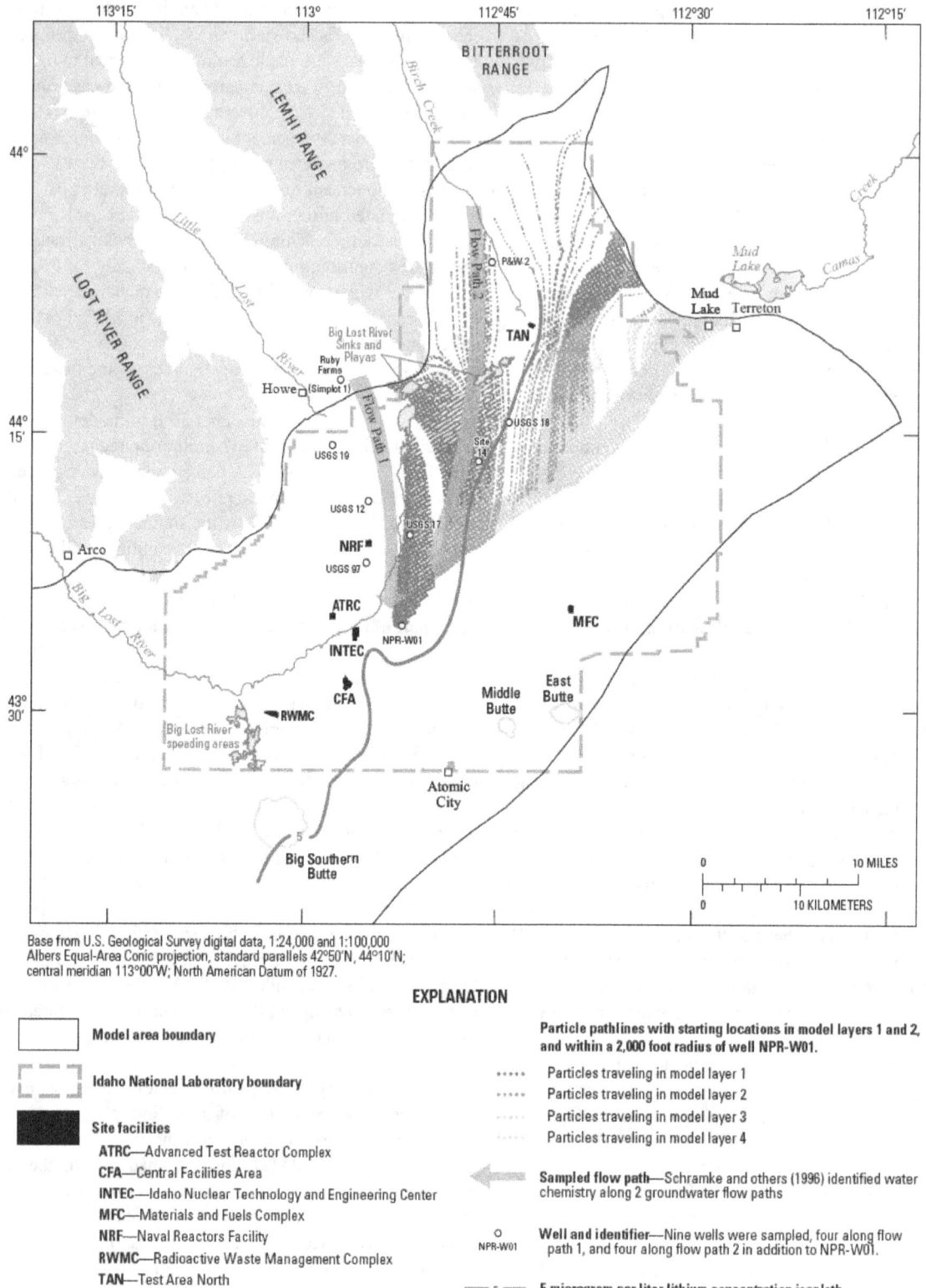

Base from U.S. Geological Survey digital data, 1:24,000 and 1:100,000
Albers Equal-Area Conic projection, standard parallels 42°50'N, 44°10'N;
central meridian 113°00'W; North American Datum of 1927.

EXPLANATION

☐	Model area boundary
☐	Idaho National Laboratory boundary
■	Site facilities

ATRC—Advanced Test Reactor Complex
CFA—Central Facilities Area
INTEC—Idaho Nuclear Technology and Engineering Center
MFC—Materials and Fuels Complex
NRF—Naval Reactors Facility
RWMC—Radioactive Waste Management Complex
TAN—Test Area North

Particle pathlines with starting locations in model layers 1 and 2, and within a 2,000 foot radius of well NPR-W01.

····· Particles traveling in model layer 1
····· Particles traveling in model layer 2
····· Particles traveling in model layer 3
····· Particles traveling in model layer 4

Sampled flow path—Schramke and others (1996) identified water chemistry along 2 groundwater flow paths

○ NPR-W01   **Well and identifier**—Nine wells were sampled, four along flow path 1, and four along flow path 2 in addition to NPR-W01.

═ 5 ═   **5 microgram per liter lithium concentration isopleth**

Figure 17.   Backward particle-tracking pathlines indicating the sources of groundwater at well NPR-W01, Idaho National Laboratory, Idaho.

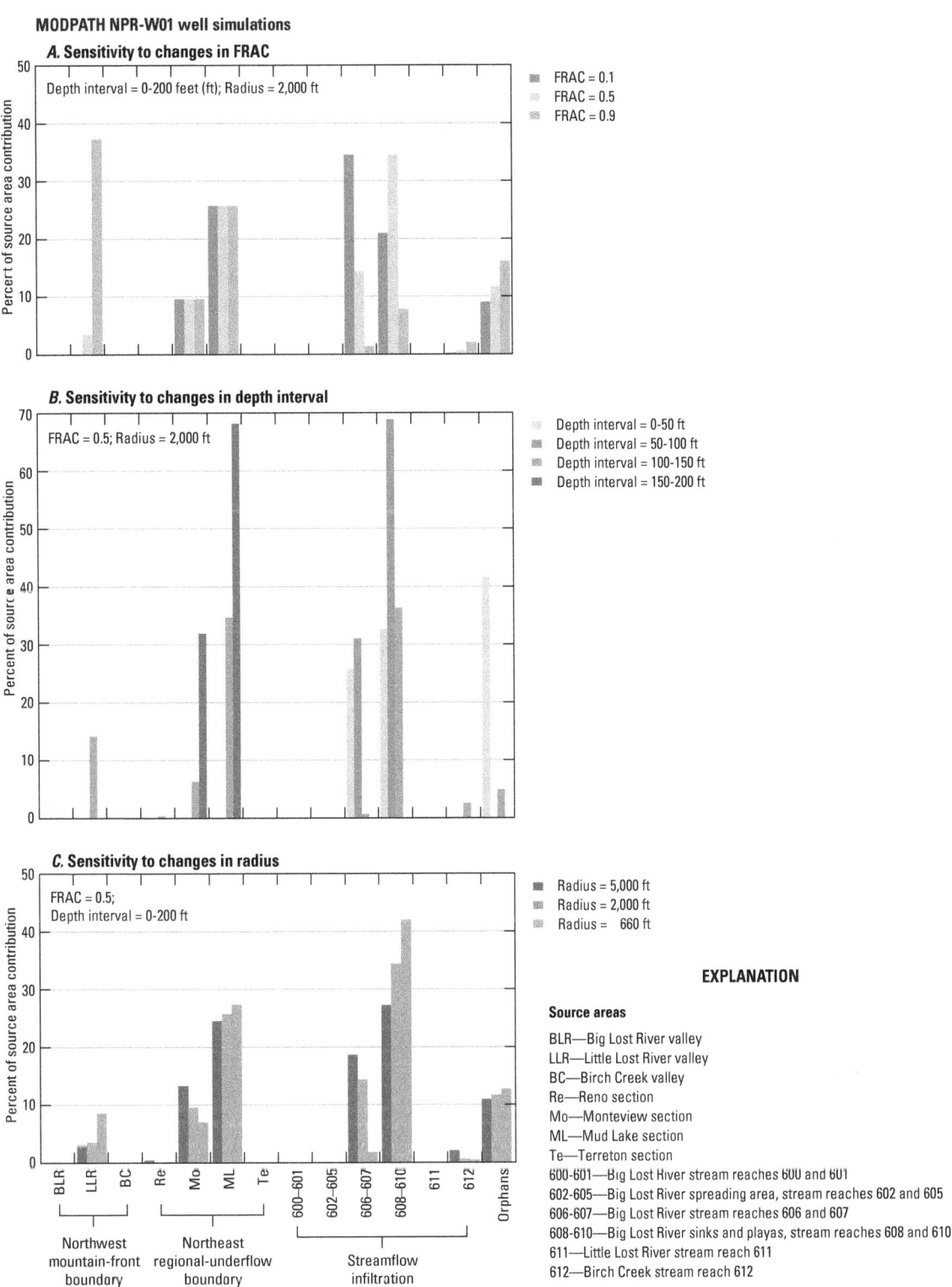

Figure 18. Sensitivity of backward particle-tracking results to changes in (*A*) FRAC, the fraction of the total outflow from a cell contributed by an internal source and used to identify weak-source cells; (*B*) Depth interval—the depth interval in which particles are located at their release locations, and (*C*) Radius, the aerial radius of the cylindrical volume defining particle release locations, Idaho National Laboratory and vicinity, Idaho. Results are shown as the percentage of source area contribution from tributary valleys (BLR, LLR, BC), streamflow-infiltration recharge (stream reaches 600 to 612), and regional aquifer water for particles released in the vicinity of the NPR-W01 well (Re, Mo, ML, Te).

Table 7.   Summary of backward particle-tracking results showing the percentage of source area contributions of tributary valley, streamflow-infiltration recharge, and regional aquifer water for particles released within the vicinity of well NPR-W01, Idaho National Laboratory and vicinity, Idaho.

[Locations of source areas are shown in figure 3. Orphans are identified as particles that did not terminate in one of the specified source areas. **Water composition:** the water composition based on geochemical mass balance and mixing models (Shramke and others, 1996). **FRAC:** the fraction of the total outflow from a cell contributed by an internal source; used to identify weak-source cells. **Depth interval:** the depths of released particles are distributed randomly in this depth interval, in feet below the top of model layer 1. **Radius:** the aerial radius of the cylindrical volume defining particle release locations, in feet. **Base case:** FRAC = 0.5; Depth interval = 0–200 feet; and Radius = 2,000 feet. **Abbreviations:** ft, foot; –, not available; NA, not applicable]

Simulated groundwater source area	Water composition	Base case	FRAC		Depth interval in feet				Radius in feet	
			0.1	0.9	0–50	50–100	100–150	150–200	5,000	660
Northwest mountain-front boundary										
Big Lost River valley (BLR)	0	0.0	0.0	0.0	0.0	0.0	0.0	0.0	0.0	0.0
Little Lost River valley (LLR)	>14	3.6	0.0	37.3	0.0	0.1	14.1	0.0	3.0	8.5
Birch Creek valley (BC)	0	0.0	0.0	0.0	0.0	0.0	0.0	0.0	0.0	0.0
Northwest mountain-front subtotal	**>14**	**3.6**	**0.0**	**37.3**	**0.0**	**0.1**	**14.1**	**0.0**	**3.0**	**8.5**
Northeast regional-underflow boundary										
Reno Ranch section (Re)	–	0.1	0.1	0.1	0.0	0.0	0.4	0.0	0.4	0.2
Monteview section (Mo)	–	9.5	9.5	9.5	0.0	0.0	6.3	31.9	13.3	7.0
Mud Lake section (ML)	–	25.7	25.7	25.7	0.0	0.0	34.7	68.1	24.5	27.3
Terreton section (Te)	–	0.0	0.0	0.0	0.0	0.0	0.0	0.0	0.0	0.0
Northeast regional-underflow subtotal	[1]**17**	**35.3**	**35.3**	**35.3**	**0.0**	**0.0**	**41.4**	**100.0**	**38.2**	**34.5**
Water table boundary										
Streamflow infiltration										
Big Lost River infiltration										
Stream reach 600–601	–	0.0	0.0	0.0	0.0	0.0	0.0	0.0	0.0	0.0
Stream reach 602–605	–	0.0	0.0	0.0	0.0	0.0	0.0	0.0	0.0	0.0
Stream reach 606–607	–	14.4	34.5	1.4	25.8	31.0	0.7	0.0	18.6	1.8
Stream reach 608–610	–	34.4	21.0	7.8	32.6	68.9	36.3	0.0	27.2	42.0
Little Lost River infiltration										
Stream reach 611	0	0.0	0.0	0.0	0.0	0.0	0.0	0.0	0.0	0.0
Birch Creek infiltration										
Stream reach 612	0	0.6	0.2	2.0	0.0	0.0	2.6	0.0	2.1	0.5
Streamflow-infiltration subtotal	[2]**<69**	**49.4**	**55.7**	**11.2**	**58.4**	**99.9**	**39.6**	**0.0**	**47.9**	**44.3**
Orphans	**NA**	**11.7**	**9.0**	**16.1**	**41.6**	**0.1**	**4.9**	**0.0**	**10.9**	**12.7**

[1] May contain a small percentage of alluvial aquifer underflow from the Birch Creek valley.

[2] Includes alluvial aquifer underflow from the Little Lost River valley.

In the base-case particle tracking simulation, the large proportion of regional aquifer water predicted by the model (35.3 percent) is not consistent with Schramke's analysis (17 percent). These large source-area contributions from the northeast regional-underflow boundary may result from particle release depths (0–200 ft below the water table) that far exceed the well's open interval (from about the water table to 47 ft below the water table). On the other hand, base-case particle release depths may adequately integrate model results obtained at different spatial scales where water samples analyzed by Schramke that were collected at a spatial scale much smaller than the resolution of the layered-model grid, may not. The large contribution from the northeast regional-underflow boundary may then indicate an overprediction by the model that results from large contrasts in hydraulic conductivity between hydrogeologic zones. For example, $K_{h,22}$ (4,780 ft/d)—much larger than $K_{h,2}$ (384 ft/d)—facilitates the diversion of regional aquifer water entering the aquifer along the Mud Lake (ML) section of the northeast boundary from zone 2 into zone 22 and likely explains much of the overprediction of Type B water in the vicinity of the NRP-W01 well and Type B water primarily within model layer 2 (Case 4). Note that underflow from the ML section represents 25.7 percent of the total source-area contributions predicted by the model (or 72.8 percent of all Type B water).

Model-derived estimates of the source area contributions from streamflow-infiltration recharge (49.4 percent) were consistent with the independently derived estimates based on relative percentages of source water chemistry (less than 69 percent). The small percentage of tributary valley water predicted by the model (3.6 percent) is not, however, consistent with Schramke's analysis (greater than 14 percent). In the vicinity of the NRP-W01 well, the ratio of Little Lost River valley underflow contributions to streamflow-infiltration recharge contributions is dependent on FRAC. For example, an increase in FRAC from 0.5 to 0.9 results in a 33.7 percent increase in source area contribution from the Little Lost River valley, a 38.2 percent decrease in contributions from streamflow infiltration, and no change in contribution from the northeast regional-underflow boundary. Small increases in FRAC produce model-derived source area contributions that are in better agreement with independently derived estimates; however, these estimates are believed to be insufficient for constraining FRAC estimates.

The simulation's underpredicted source area contribution from the Little Lost River valley also may indicate a short-circuiting of underflow from the Little Lost River valley to an area of high hydraulic conductivity (figs. 2 and 15B). Inflows entering zone 11 ($K_{h,11}$ = 227 ft/d) from the Little Lost River valley move a relatively short distance before they are diverted towards zone 1 ($K_{h,1}$ = 11,700 ft/d) and away from the vicinity of well NPR-W01 (Case 1). The model's representation of the hydraulic conductivity in zone 1 may be too large.

An Evaluation of the Estimated Linear Velocity of the Young Fraction of Groundwater at Multiple Wells

In this analysis, independently derived estimates of the average linear velocity of groundwater within the upper 100 ft of the aquifer are compared to model-derived estimates of groundwater velocity. Geographic source areas used to calculate the average linear velocity of groundwater for 23 locations are based on estimated model ages from Busenberg and others (2001) and from independently derived estimates from lithium concentration in groundwater (table 8). Figure 19A gives the flow directions based on an estimated source area for each well. Independently derived estimates of the age of the young fraction, defined as groundwater that has been in contact with the atmosphere during the last 60 years as a result of natural recharge, irrigation return flow, or industrial wastewater disposal, are based on $^3H/^3He$ model ages. The methodologies used to identify and date the young fraction are described in Busenberg and others (2001). The age of the young fraction represents an average age at the time of sampling and is not a constant. Both the fractional percent of young water and its integrated average age depend on the recharge mechanism(s), frequency, timing, and magnitude of recharge events prior to sampling. Thus, the young fraction represents a mixture of waters of different ages that could vary from 0 to about 60 years. Age variability likely is greatest in areas where the unsaturated zone is thin, and (or) where recharge is concentrated; for example, in areas affected by streamflow-infiltration recharge, irrigation return flow, wastewater disposal, and runoff accumulation into closed drainage basins with rapid, focused recharge.

The independently derived estimate of the average linear velocity is calculated for each well by dividing the linear distance between the well and its interpreted source area by the age of the young fraction of groundwater (table 8). For example, a 45,260-ft linear distance separates well USGS 109 from its interpreted source area, the INTEC disposal well; the estimated age of the young fraction of groundwater for this well is 7,300 days; therefore, the average linear velocity is calculated as 6.2 ft/d. For wells USGS 1 and USGS 100, average linear velocities were calculated assuming a source area from the Mud Lake–Terreton area (Busenberg and others, 1993, p. 30). Evidence of rapid, focused recharge between MFC and Atomic City (Busenberg and others, 2001, p. 43) indicates that source areas for wells USGS 1 and 100 are much closer than originally thought; therefore, average linear velocity estimates at these wells should be considered as upper limits (table 8). The average linear velocities for all 23 wells ranged from 1.6 to less than 21 ft/d, with the smallest and largest velocities at wells USGS 102 and USGS 100, respectively, an arithmetic mean of 8.0 ft/d and a standard deviation of 4.3 ft/d.

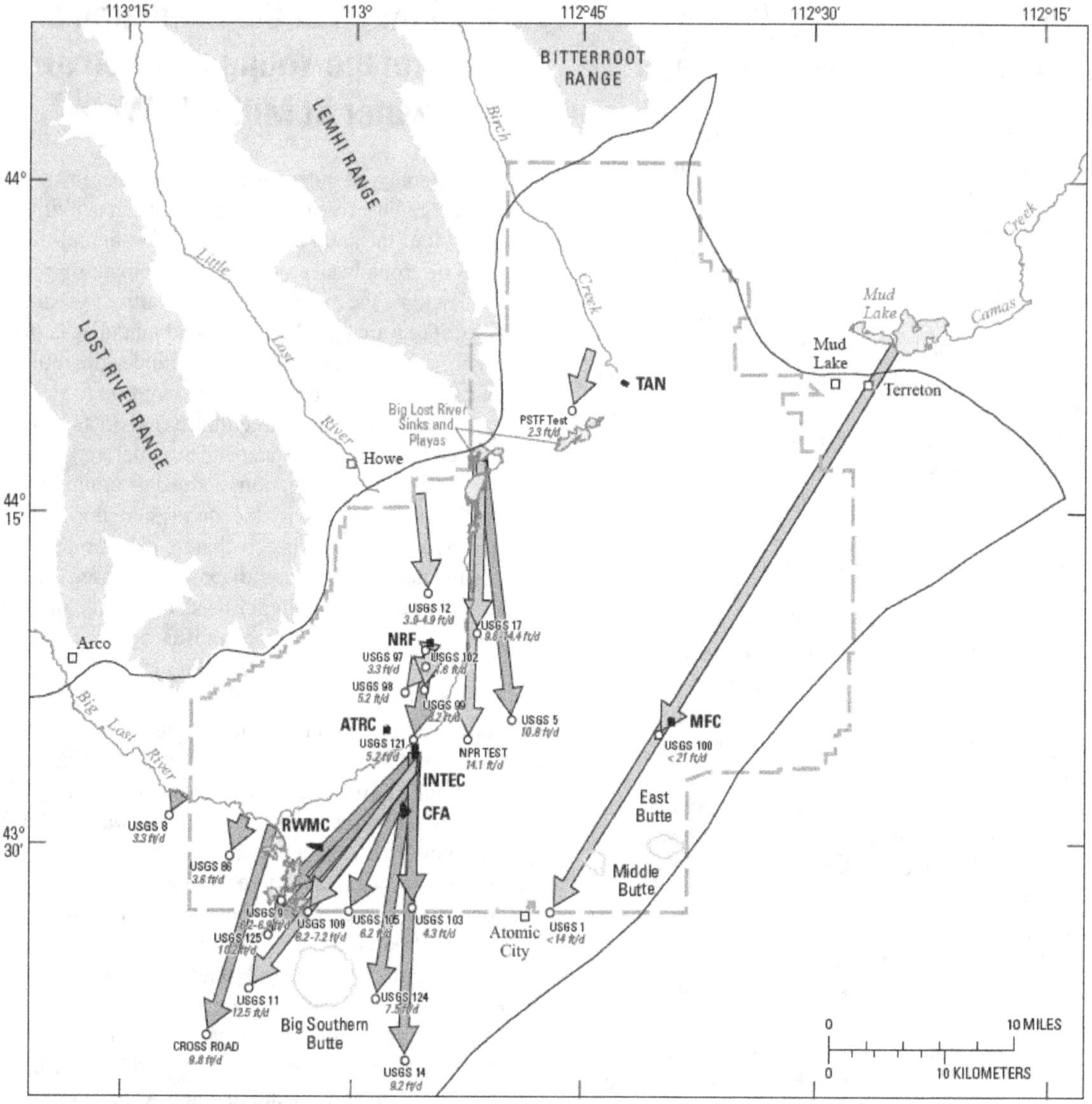

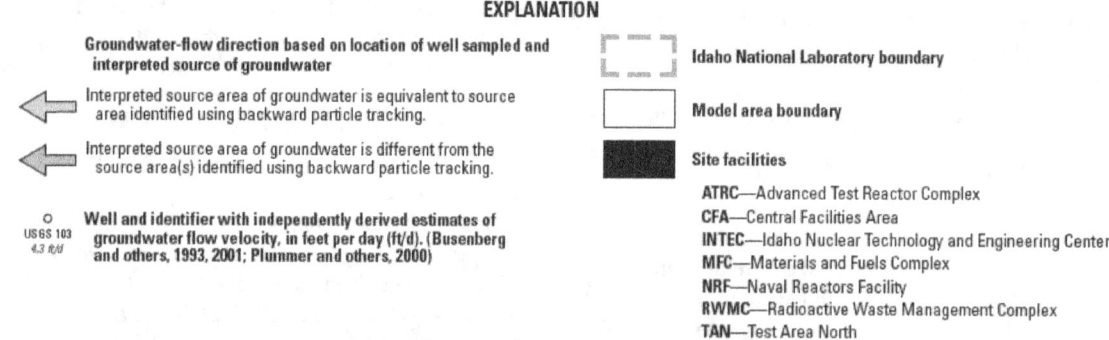

A. Independently derived estimates of groundwater velocity

Figure 19.   Selected wells (*A*) independently derived estimates of groundwater velocity, and (*B*) model derived estimates of groundwater velocity, Idaho National Laboratory and vicinity, Idaho.

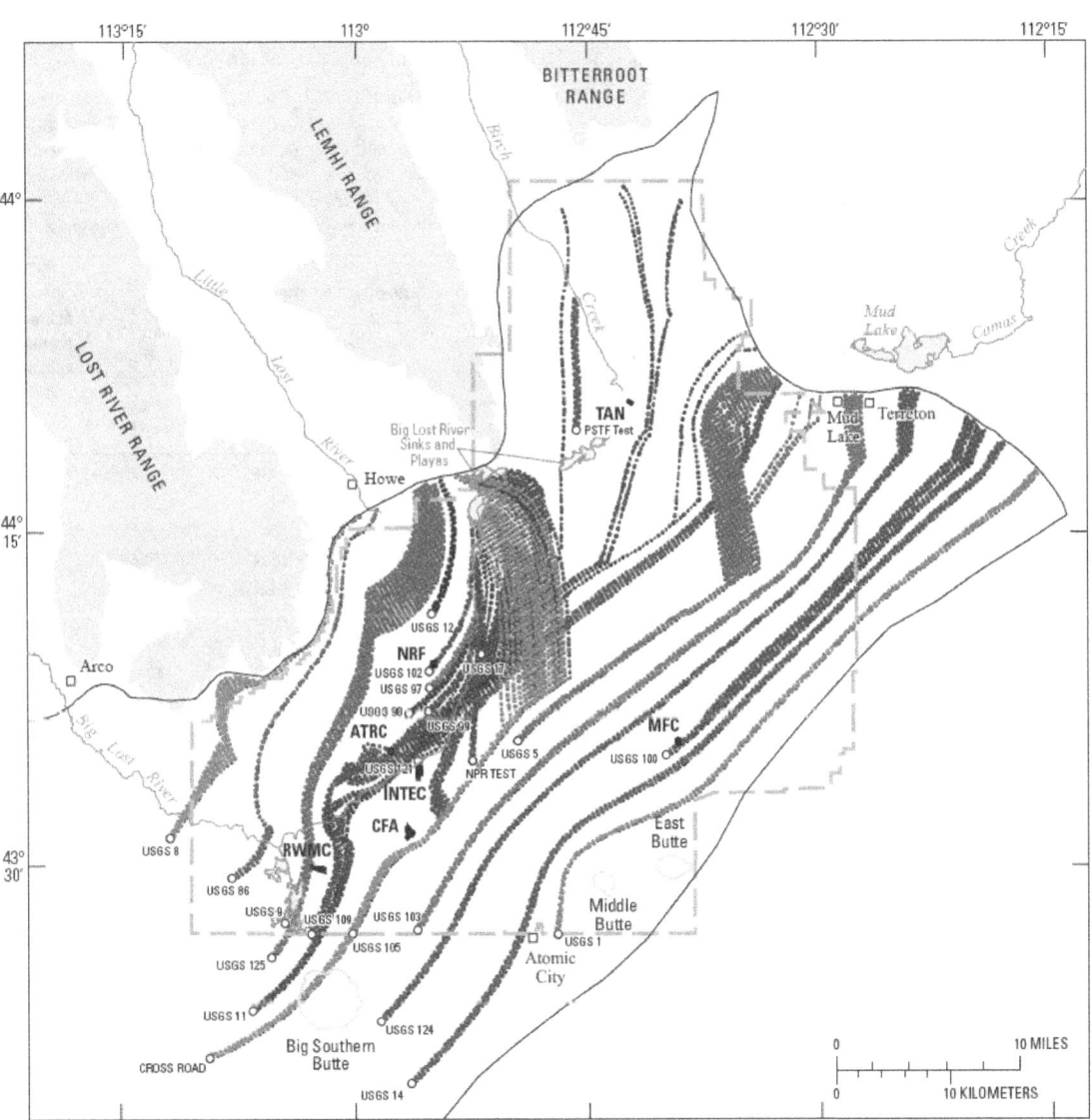

Base from U.S. Geological Survey digital data, 1:24,000 and 1:100,000
Albers Equal-Area Conic projection, standard parallels 42°50'N, 44°10'N;
central meridian 113°00'W; North American Datum of 1927.

B. Model-derived estimates of groundwater velocity

EXPLANATION

Particle pathlines with starting locations in model layer 1—
Particle color represents the well at which the particle was
located at release

○   Well and identifier—Names are local well names
USGS 11

Idaho National Laboratory boundary

Model area boundary

Site facilities

ATRC—Advanced Test Reactor Complex
CFA—Central Facilities Area
INTEC—Idaho Nuclear Technology and Engineering Center
MFC—Materials and Fuels Complex
NRF—Naval Reactors Facility
RWMC—Radioactive Waste Management Complex
TAN—Test Area North

Figure 19.—Continued

Table 8.   Average groundwater velocities calculated using independently derived estimates of travel time and backward particle tracking runs, between selected wells and source areas, Idaho National Laboratory and vicinity, Idaho.

[**Local name:** Local well identifier used in this study. **Water type:** Based on criteria used by Olmsted (1962). The methodologies used to identify the age of the young fraction of groundwater are described in Busenberg and others (2001). Average linear flow velocities for the young fraction of groundwater are calculated based on tritium/helium-3 model ages. **Geographic source area:** tv, tributary valley; si, surface infiltration; ra, regional aquifer; LLR, Little Lost River valley; Re, northeast boundary Reno section; Mo, northeast boundary Monteview section; ML, northeast boundary Mud Lake section; Te, northeast boundary Terreton section; 606–607, between streamflow-gaging stations 504 and 506; 608–610, within the Big Lost River sinks and playas; 612, within Birch Creek downstream from the model boundary. **Abbreviations:** Li, lithium; µg/L, micrograms per liter; yr, year; ft/d, foot per day; –, not available; NA, not applicable]

Local name	Water type	Li concentration (µg/L)	Geographic source area based on Li concentration	Geographic source area based on isotopic and dissolved gas data	Tritium/ helium-3 model age of the young fraction of groundwater (yr)	Average linear velocity based on model ages (ft/d)	Geographic source area based on particle tracking	Averge travel time for particle to reach source area (yr)	Average particle velocity (ft/d)	Velocity difference for equivalent source areas (ft/d)
USGS 1	B	18.0, 22	ra	[1]Mud Lake–Terreton	32	[2]<14	Te	8.4	62.9	> 49
USGS 5	A	2.0	tv, si	BLR Sinks	16.5 ± 0 5	10.8	ML	27.1	15.2	NA
				BLR Sinks	16.3 ± 0 3	10.8	NA	NA	NA	NA
USGS 8	A	1 3, 6	tv, si	BLR channel	8.4 ± 0 2	3 3	LLR	33.4	9.6	NA
USGS 9	A/B	3 2, 3 3, <4	tv, si	INTEC disposal well	21.3 ± 0.6	6 9	LLR	11.9	30.0	NA
				INTEC disposal well	22.7 ± 0.4	6 2	NA	NA	NA	NA
USGS 11	A	2 1, 2.1	tv, si	INTEC disposal well	17.3 ± 0 3	12 5	LLR	66.0	8.3	NA
							ML	42.6	16.2	NA
							Mo	44.5	15.4	NA
							Re	56.4	13.3	NA
							606–607	5.9	35.4	27.0
							608–610	61.0	7.7	NA
							612	58.9	12.9	NA
USGS 12	B/C	2.7	tv, si	LLR underflow	2.9 ± 0.4	4 9	LLR	10.5	10.0	5.6
				LLR underflow	4.5 ± 0.4	3 9	NA	NA	NA	NA
USGS 14	B	24.3	ra	INTEC disposal well	27.3 ± 0 5	9 2	Te	10.5	64.8	NA
USGS 17	A/B/C	1.4, 4	tv, si	BLR Sinks	16.1 ± 0 3	9.8	608–610	13.1	6.2	–5.9
				BLR Sinks	11.1 ± 0 3	14.4	NA	NA	NA	NA
USGS 86		2 3	tv, si	BLR channel	12.1 ± 0 5	3.6	LLR	18.9	16.8	NA
USGS 97	B/C	2.6	tv, si	NRF	6.3 ± 0 5	3 3	608–610	23.2	6.2	2.9
USGS 98	A	2 5	tv, si	NRF	6.7 ± 1 3	6.6	608–610	30.9	5.5	–1.1
USGS 99	A	2 5	tv, si	NRF	3.9 ± 0 2	8 2	608–610	29.2	5.5	–2.7
USGS 100	B	22.6, 23.4	ra	[1]Mud Lake–Terreton	14.4–15 2	[2]<21	Te	6.7	52.1	>31
USGS 102	A	2 9	tv, si	NRF ditch	5.7 ± 0 2	1.6	608–610	19.0	6.8	5.2
USGS 103	A/B/C	6 9	ra	INTEC disposal well	26.1 ± 0.4	4 3	ML	9.3	59.3	NA
USGS 105	A/B	2 5	tv, si	INTEC disposal well	–	[3]6 2	LLR	38.7	11.6	NA
							ML	32.8	17.5	NA
							608–610	32.0	12.4	NA
USGS 109	A	3, 2.7	tv, si	INTEC disposal well	20.0 ± 0.4	6 2	LLR	52.4	9.1	NA
				INTEC disposal well	17.7 ± 0.4	7 2	606–607	26.9	9.2	2.0
							608–610	48.0	8.7	NA
USGS 121	–	5	tv, si, ra	NRF	15.5 ± 0.6	5 2	606–607	1.4	2.1	NA
USGS 124	A	6 9, 6.7	ra	INTEC disposal well	23.7 ± 0 1	7 5	ML	10.3	62.2	NA
				INTEC disposal well	23.6 ± 0 5	7 5	NA	NA	NA	NA
USGS 125	A	3 2, 3.1	tv, si	INTEC disposal well	17.0 ± 0 3	10 2	LLR	15.4	26.0	NA
Crossroads Well	–	4	tv, si	BLR spreading areas	[4]13.1 ± 0.4	9.8	LLR	41.9	14.3	NA
							608–610	34.4	15.5	NA
NPR Test	A	2, 2.2, 4	tv, si	BLR Sinks	13.9 ± 0.4	14 1	608–610	36.3	4.5	–9.6
PSTF Test	A	1.8	tv, si	Birch Creek Playa	9.3 ± 2 2	2 3	612	11.5	8.4	6.1

[1] Busenburg and others (1993, p. 30).
[2] Less than sign added to reflect evidence of rapid focused recharge between the Materials and Fuels Complex and Atomic City (Busenberg and others, 2001, p. 43).
[3] Busenburg and others (2001, fig. 25).
[4] Plummer and others (2000).

Model-derived estimates of the average groundwater velocity are based on the results of backward particle-tracking simulations at 23 well locations involving the release of 63 particles from each well, distributed uniformly along three evenly spaced horizontal planes at depths of 25, 50, and 75 ft below the water table boundary (the top of model layer 1) (fig. 19B). These simulations were limited to wells open only to model layer 1; consequently, particle simulations were restricted to model layer 1 and comparison wells were chosen that penetrate only the upper 100 ft of the aquifer. For each well, particle pathlines were grouped according to their geographical source area (fig. 3), an average particle velocity (average groundwater velocity) calculated for each pathline in a group, and the average of these velocities reported in table 8. Model-derived average groundwater velocities for all 23 wells ranged from 2.1 to 64.8 ft/d, with the smallest and largest velocities at wells USGS 121 and USGS 14, respectively, an arithmetic mean of 19.5 ft/d and a standard deviation of 18.4 ft/d.

A direct comparison between model-derived and independently derived estimates of the groundwater velocity at a well is possible where estimates are based on equivalent geographical source areas; 12 of the 23 wells met this criterion and include wells USGS 1, 11, 12, 17, 97, 98, 99, 100, 102, 109, NPR Test, and PSTF Test. Independently derived estimates of the average linear velocity are dependent on the age of the young fraction of groundwater; therefore, comparisons were limited to those source areas having a surface-water component. Differences between velocity estimates for equivalent source areas ranged from –9.6 (68 percent error) to 48.9 ft/d (349 percent error) with the smallest and largest velocity estimates at well NPR Test and well USGS 1, respectively. Agreement between velocity estimates was good at wells (USGS 12, 17, 97 98, 99, 102, NPR Test, and PSTF Test) with travel paths located in areas of sediment-rich rock (zones 11, 22, 44); velocity differences at these wells ranged from -9.6 to 6.1 ft/d with the smallest and largest velocity differences at NPR Test and PSTF Test, respectively. For wells USGS 1 and USGS 100 with travel paths located in areas of sediment-poor rock (zone 1), velocity differences were greater than 31 ft/d and simulated velocities are 2.5 to 4.5 times larger than independently derived estimates at well USGS 1 (less than 14 ft/d) and well USGS 100 (less than 21 ft/d). The velocity difference was also large (27.0 ft/d) at well USGS 11; its travel path is located primarily in an area of sediment-poor rock (zone 1). The USGS 109 travel path is located in areas of both sediment-poor and sediment-rich rock (zone 1 and zone 11), and the velocity difference at this well is small (2.0 ft/d).

The overprediction by the model of groundwater velocities in sediment-poor rock may be attributed to large contrasts in hydraulic conductivity and a large, model-wide estimate of vertical anisotropy. A $K_{h,1}$ that is much larger than $K_{h,2}$ in the east-central part of the model area inhibits both lateral and downward flow from zone 1 into zone 2 because the horizontal (11,700 ft/d) and vertical (0.79 ft/d) hydraulic conductivities of zone 1 are much larger than the horizontal (384 ft/d) and vertical (0.03 ft/d) conductivities

of zone 2 (Case 5). These relations can account for the very high velocities (greater than 100 ft/d) in model layers 1 and 2 (figs. 15B and D) where regional aquifer groundwater in zone 1 flows upward and over zone 2 where it subcrops across the tops of model layers 2 and 3 (fig. 2) (Case 5).

The effect of a very large vertical anisotropy (14,800) is a very small vertical hydraulic conductivity (K_z) of all hydrogeologic zones. Thus, significant vertical flow requires a very large vertical head gradient. Hydraulic head measurements in five MLMS wells (USGS 103, 105, 108, 132, and 135) (figs. 1 and 16) located in sediment-poor rock (zones 1–3) along and near the southern boundary of the INL indicate that vertical changes in measured heads are small, varying by only 0.3 to 1.4 ft from the water table to depths of 734 ft below the water table. Modeled vertical head changes are very large in comparison, resulting in simulated vertical head gradients in the upper 800 ft of the aquifer that are about 46 to 239 times larger than the measured vertical head gradients in the MLMS wells USGS 103 (8.3×10^{-4} feet/foot [ft/ft]), 105 (5.5×10^{-4} ft/ft), 108 (1.5×10^{-3} ft/ft), 132 (4.8×10^{-4} ft/ft), and 135 (2.9×10^{-3} ft/ft) (Brian Twining, U.S. Geological Survey, written commun., 2012). Evidence of vertical interconnectedness among fracture sets in the sediment-poor rock is indicated by temperature measurements in the five MLMS wells (USGS 103, 105, 108, 132, and 135); temperature profiles in these wells are dominated by convective heat transfer, where fluid flow through the fractured media significantly altered the geothermal field (Brian Twining, U.S. Geological Survey, written commun., 2012). The overprediction of vertical head gradients by the model and evidence of vertical flow among fracture sets indicates that a very small K_z or large vertical anisotropy is not representative of sediment-poor rock. These low vertical conductivities can account for the very large simulated groundwater velocities in areas of sediment-poor rock; in these areas an underprediction of volumetric flux in the vertical direction results in an overprediction of flux in the horizontal direction.

The overprediction by the model of groundwater velocities in sediment-poor rocks could also result from an underprediction of effective porosity in hydrogeologic zone 1 ($n_1 = 0.07$). The functional relation between average groundwater velocity and effective porosity is shown in equation (6). In order to get the simulated average velocity of well USGS 1 (62.9 ft/d) commensurate with its independently derived estimate (less than 14 ft/d) requires an n_1 that is about 4.5 times larger (0.32). For well USGS 100, a 2.5 times larger n_1 value (0.18) is required to decrease the simulated average velocity (52.1 ft/d) to a value commensurate with the independently derived estimate (less than 21 ft/d). The revised estimates of n_1 are well outside the 95-percent confidence interval for specific yield in hydrogeologic zone 1 (0.068–0.077), and in the case of n_1 equal to 0.32, outside the expected interval for specific yield in zone 1 (0.01–0.30) (table 2), thus indicating that modeled values of effective porosity do not account for the very large simulated groundwater velocities in areas of sediment-poor rock.

Summary and Conclusions

Three backward particle-tracking simulations were used to (1) trace the sources of groundwater in the model area back to the point where groundwater crosses an inflow boundary, and (2) estimate groundwater velocities within the model area. The simulations used the U.S. Geological Survey three-dimensional model of steady-state groundwater flow in the west-central part of the eastern Snake River Plain (ESRP) aquifer and a modified version of the particle-tracking program MODPATH. Results of these simulations were used to compare model-derived to independently derived estimates of groundwater source areas and velocities.

Groundwater source areas used to evaluate the source area simulations were defined by differences in the (1) cationic and anionic composition of groundwater that are based primarily on the reactive percentages of sodium plus potassium relative to that of calcium plus magnesium, and the reactive percentages of bicarbonate plus carbonate relative to other cationic components, and (2) concentrations of the trace elements lithium (Li), boron (B), and fluoride, and concentrations of the dissolved gas helium.

The trace elements Li and B, in conjunction with the water-typing classifications Type A, Type B, and Type C (Olmsted, 1962) and Type I, Type II, and Type III (Busenberg and others, 2001), were used to define a boundary separating groundwater within the upper 200 feet (ft) of the aquifer (derived primarily from tributary valley underflow and streamflow-infiltration recharge) from groundwater that is derived primarily from regional aquifer underflow. The position of this line was defined by a sharp break in Li concentrations in a rank-ordered plot of Li and B concentrations and B to Li (B/Li) concentration ratios from 87 groundwater and thermal spring sampling sites in the study area. Groundwater with less than about 5 micrograms per liter (μg/L) Li is derived primarily from tributary valley underflow along the northwest mountain-front boundary and from streamflow-infiltration recharge in the Big Lost River, Little Lost River, and Birch Creek. This water is equivalent to the Type A water of Olmsted (1962) and the Type I water of Busenberg and others (2001). Groundwater with Li concentrations greater than about 5 μg/L is derived primarily from regional aquifer underflow along the northeast boundary, from streamflow-infiltration recharge in the Camas Creek and Mud Lake, and agricultural return flow in the Mud Lake and Terreton areas north of the northeast model boundary. This water is equivalent to the Type B water of Olmsted (1962) and the Type II and Type III waters of Busenberg and others (2001). The 5 μg/L Li line extends from the northern to the southern end of the model area, is parallel to the regional direction of groundwater flow, and closely approximates the position of the line that Olmsted defined separating Type A from Type B waters.

A two-component mixing model, using Li concentrations and B/Li concentration ratios, was developed to describe mixing of tributary valley and regional aquifer groundwater

east of the 5 μg/L line. Mean values of Li concentrations and B/Li concentration ratios were used to represent the end-member contributions from tributary valley groundwater and regional aquifer groundwater. Using mean values, the mixed zone is bounded by Li concentrations of 2.9 μg/L (standard deviation [σ]=0.9) for tributary valley water and 20.8 μg/L (σ=12.3) for regional aquifer water, and B/Li concentration ratios of 2 (σ=1) for regional aquifer water and 8 (σ=2) for tributary valley water. The resulting model is hyperbolic in form and highly nonlinear. Although the model fit to the Li and B/Li data is very good (coefficient of determination [R^2] =0.71), the reliability of the end-member estimate for the Li concentration of regional aquifer water is very uncertain. This uncertainty is reflected in the range (5.2 to 72.3 μg/L) and large standard deviation (12.3 μg/L) for Li concentrations from regional aquifer water relative to its mean value (20.8 μg/L) and in the strong northeast to southwest spatial trend in Li isopleths east of the 5 μg/L Li line. An alternative two-component mixing model was developed using a curve-fitting regression technique to identify the end member variables with the largest uncertainty. Four regression-based models with equally valid results were identified (R^2=0.78 and root-mean-square error [RMSE]=1.47). Model 1 imposes the least restrictive bounds on the Li concentration in the tributary valley ([Li]$_{tv}$=2.9 μg/L) and B/Li concentration ratio in the regional aquifer (B/Li)$_{ra}$=2), and Model 4 imposes the most restrictive bounds on these end members ([Li]$_{tv}$=3.8 μg/L and (B/Li)$_{ra}$=3). Using the bounding criteria for mixed water based on the regression-based models, the interpolated contour maps of Li concentration and B/Li concentration ratio, and the chemical characteristics of groundwater in wells located in the vicinity of and east of the 5 μg/L Li line, the mixing zone, within the upper 200 ft (model layers 1 and 2) of the aquifer, is described as relatively narrow and is probably no more than about 1 to 2 miles wide along its dominant northeast to southwest trend.

Backward particle-tracking was used to identify the contributing source areas of groundwater and to estimate groundwater velocities within each model layer at each particle release location. Particles were released in an aerially uniform pattern covering the entire model domain. Resulting particle pathlines indicate that the source of groundwater in the upper 200 ft (model layers 1 and 2) of the aquifer is dominated by (1) inflow from the tributary valleys, (2) streamflow infiltration west of the 5 μg/L Li line, and by (3) regional aquifer inflows east of this line. In successively deeper model layers the contribution of inflow from the tributary valleys and surface-water sources gradually diminishes with the thickening of the aquifer toward the east. Simulated velocities were commensurate with the aquifer's hydraulic conductivity distribution and gradually decreased with depth. In the upper model layers, low-velocity tributary valley and regional aquifer underflow water that enters from the northern boundaries travels in a southerly direction until it merges with the high-velocity regional water that enters from the northeast boundary and travels in a southwesterly direction. A sharp refraction of flowlines occurs at the intersection of low- and high-velocity flows; this refraction results from a contrast in hydraulic conductivity of almost 2 orders of magnitude between hydrogeologic zone 1 (11,700 ft/d) and zone 11 (227 ft/d). This contrast diminishes in the lower model layers where velocity differences between tributary valley and regional water are relatively small. In the upper two model layers the simulated position of the line separating groundwater derived from tributary valley underflow and streamflow-infiltration recharge sources (Type A water) from regional aquifer underflow sources (Type B water) was compared to the 5 µg/L Li line. These comparisons indicate several shortcomings in the way the model represents flow in the aquifer. The model simulates too much eastward movement of tributary valley underflow and streamflow-infiltration recharge in the north-central part of the model area and not enough of this flow in the central part of the model area. Inconsistencies between model-derived flows and the independently derived estimate of the interface between Type A and Type B waters can be attributed to large contrasts in hydraulic conductivity between hydrogeologic zones.

Backward particle-tracking simulations of particles released in the vicinity of well NPR-W01 were used to compare model-derived estimates of the sources and quantities of groundwater to independently derived estimates based on chemistry data that were previously published about mixed groundwater at well NPR-W01. Resulting particle pathlines indicate that groundwater at well NPR-W01 represents a mixture of water comprising 3.6 percent Little Lost River underflow, 35.3 percent regional aquifer underflow, and 49.4 percent streamflow-infiltration recharge, with orphans accounting for the remaining 11.7 percent. Source area contributions from the Big Lost River make up the bulk of surface-water recharge at 48.8 percent, and the remaining 0.6 percent comes from Birch Creek recharge. Geochemical mass-balance and mixing models indicate that the composition of water collected from well NPR-W01 is greater than about 14 percent underflow from the Little Lost River valley, less than 69 percent streamflow-infiltration recharge from the Big Lost River (this includes alluvial-aquifer underflow from the Little Lost River valley), and 17 percent inflow from the northeast boundary (that may contain a small percentage of underflow from the Birch Creek valley). The large proportion of regional aquifer water predicted by the model (35.3 percent) is inconsistent with the geochemical analysis (17 percent) and may be caused by the large hydraulic conductivity contrasts between hydrogeologic zones simulated in the model. The model's underprediction of the source area contribution from the Little Lost River valley may indicate a short-circuiting of underflow from the Little Lost River valley to an area of high hydraulic conductivity.

Tritium/helium-3 based estimates of the age of the young fraction of groundwater were used to compare model-derived

estimates of groundwater velocities to independently derived estimates within the upper 100 ft of the aquifer. The age of the young fraction, defined as groundwater that has been in contact with the atmosphere during the last 60 years, represents an average age at the time of sampling. Independently derived estimates of groundwater velocity for each well were calculated by dividing the linear distance between the sampled well and its source area by the age of the young fraction of groundwater. These velocity estimates represent average linear velocities. Model-derived estimates of the average groundwater velocity were based on the results of backward particle-tracking simulations at 23 well locations. Direct comparisons between model-derived and independently derived estimates of the groundwater velocity at a well were possible where estimates were based on equivalent geographical source areas; 12 of the 23 wells met this criterion. Differences between velocity estimates for equivalent source areas ranged from –9.6 ft/d (68 percent error) to 48.9 ft/d (349 percent error). Agreement between velocity estimates was good at wells with travel paths located in areas of sediment-rich rock (RMSE = 5.2 ft/d) and poor in areas of sediment-poor rock (RMSE = 26.2 ft/d); simulated velocities in sediment-poor rock were 2.5 to 4.5 times larger than independently derived estimates at well USGS 1 (less than 14 ft/d) and well USGS 100 (less than 21 ft/d). The larger simulated groundwater velocities in sediment-poor rock may be attributed to the large hydraulic conductivity contrasts between sediment-rich and sediment-poor areas, and a very large, model-wide vertical anisotropy (14,800) simulated in the model.

Uncertainties are associated with all estimates of groundwater movement in the ESRP aquifer, and no single estimation technique can be expected to provide a complete view of the flow dynamics in this complex aquifer system. Multiple estimation techniques are needed to identify a set of consistent flow features that may be used to evaluate strengths and shortcomings of the present numerical flow model. Findings in this report indicate that present steady-state simulations of advective transport may not be realistic. Because predictive models of contaminant transport are highly dependent on flow simulations, it is suggested that future groundwater modeling in the ESRP aquifer (1) remain at the sub-regional scale, (2) simulate transient flow conditions, (3) reinterpret the geologic framework in the area of high hydraulic conductivity located directly southwest of the Little Lost River valley, (4) specify separate vertical anisotropy values for areas of sediment-rich and sediment-poor rock, (5) place controls on the vertical head gradient by including head observations from wells configured with multilevel monitoring systems as part of a formal parameter estimation, (6) reconcile inconsistencies with water-type separation by including geochemical observations as part of a formal parameter estimation, and (7) better quantify the uncertainties of modeled inflows along the northwest mountain-front boundary and northeast regional-underflow boundary.

References Cited

Ackerman, D.J., Rattray, G.W., Rousseau, J.P., Davis, L.C., and Orr, B.R., 2006, A conceptual model of ground-water flow in the eastern Snake River Plain aquifer at the Idaho National Laboratory and vicinity with implications for contaminant transport: U.S. Geological Survey Scientific Investigations Report 2006-5122 (DOE/ID-22198), 62 p. (Also available at http://pubs.er.usgs.gov/usgspubs/sir/ sir20065122.)

Ackerman, D.J., Rousseau, J.P., Rattray, G.W., and Fisher, J.C., 2010, Steady-state and transient models of groundwater flow and advective transport, eastern Snake River Plain aquifer, Idaho National Laboratory and vicinity, Idaho: U.S. Geological Survey Scientific Investigations Report 2010–5123 (DOE/ID-22209), 220 p. (Also available at http://pubs.usgs.gov/sir/2010/5123.)

Bartholomay, R.C., Tucker, B.J., Ackerman, D.J., and Liszewski, M.J., 1997, Hydrologic conditions and distribution of selected radiochemical and chemical constituents in water, Snake River Plain aquifer, Idaho National Engineering Laboratory, Idaho, 1992 through 1995: U.S. Geological Survey Water-Resources Investigations Report 97–4086 (DOE/ID-22137), 57 p. (Also available at http://pubs.er.usgs.gov/usgspubs/wri/ wri974086.)

Bartholomay, R.C., Tucker, B.J., Knobel, L.L., and Mann, L.J., 2001, Radiochemical and chemical constituents in water from selected wells south of the Idaho National Engineering and Environmental Laboratory, Idaho: U.S. Geological Survey Open-File Report 2001–138 (DOE/ID-22175), 19 p. (Also available at http://pubs.er.usgs.gov/usgspubs/ofr/ ofr01138.)

Bartholomay, R.C., and Twining, B.V., 2010, Chemical constituents in groundwater from multiple zones in the eastern Snake River Plain aquifer at the Idaho National Laboratory, Idaho, 2005–08: U.S. Geological Survey Scientific Investigations Report 2010–5116 (DOE/ID -22211), 82 p. (Also available at http://pubs.er.usgs.gov/ usgspubs/sir/sir20105116.)

Busenberg, E., Plummer, L.N., Doughten, M.W., Widman, P.K., and Bartholomay, R.C., 2000, Chemical and isotopic composition and gas concentrations of ground water and surface water from selected sites at and near the Idaho National Engineering and Environmental Laboratory, Idaho, 1994–97: U.S. Geological Survey Open-File Report 00–81 (DOE/ID-22164), 46 p. (Also available at http://pubs. er.usgs.gov/usgspubs/ofr/ofr0081.)

Busenberg, E., Plummer, L.N., and Bartholomay, R.C., 2001, Estimated age and source of the young fraction of ground water at the Idaho National Engineering and Environmental Laboratory: U.S. Geological Survey Water-Resources Investigations Report 2001–4265 (DOE/ID-22177), 144 p. (Also available at http://pubs.er.usgs.gov/usgspubs/wri/ wri014265.)

Busenberg, E., Weeks, E.P., Plummer, L.N., and Bartholomay, R.C., 1993, Age dating ground water by use of chlorofluorocarbons (CCl_3F and CCl_2F_2), and distribution of chlorofluorocarbons in the unsaturated zone, Snake River Plain Aquifer, Idaho National Engineering Laboratory, Idaho: U.S. Geological Survey Water-Resources Investigations Report 93–4054, DOE/ID-22107, 47 p. (Also available at http://pubs.er.usgs.gov/usgspubs/wri/ wri934054.)

Fisher, J.C., and Twining, B.V., 2011, Multilevel groundwater monitoring of hydraulic head and temperature in the eastern Snake River Plain aquifer, Idaho National Laboratory, Idaho, 2007–08: U.S. Geological Survey Scientific Investigations Report 2010–5253, DOE/ID-22213, 62 p. (Also available at http://pubs.usgs.gov/sir/2010/5253.)

Freeze, R.A., and Cherry, J.A., 1979, Groundwater: New Jersey, Prentice-Hall Inc., 604 p.

Harbaugh, A.W., Banta, E.R., Hill, M.C., and McDonald, M.G., 2000, MODFLOW-2000, the U.S. Geological Survey modular ground-water model—User guide to modularization concepts and the ground-water flow process: U.S. Geological Survey Open-File Report 2000–92, 121 p. (Also available at http://pubs.er.usgs.gov/publication/ ofr200092.)

Knobel, L.L., Bartholomay, R.C., Tucker, B.J., Williams, L.M., and Cecil, L.D., 1999, Chemical constituents in ground water from 39 selected sites with an evaluation of associated quality assurance data, Idaho National Engineering and Environmental Laboratory and vicinity, Idaho: U.S. Geological Survey Open-File Report 99–246, DOE/ID-22159, 58 p. (Also available at http://pubs.er.usgs. gov/usgspubs/ofr/ofr99246.)

Langmuir, C.H., Vocke, R.D., Jr., Hanson, G.N., 1978, A general mixing equation with applications to Icelandic basalts: Earth and Planetary Science Letters, v. 37, p. 380-392.

Lee, S., Wolberg, G., and Shin, S.Y., 1997, Scattered data interpolation with multilevel B-splines: IEEE Transactions on Visualization and Computer Graphics, v. 3, no. 3, p. 229–244.

Lindholm, G.F., Garabedian, S.P., Newton, G.D., and Whitehead, R.L., 1988, Configuration of the water table and depth to water, spring 1980, water-level fluctuations, and water movement in the Snake River Plain regional aquifer system, Idaho and eastern Oregon: U.S. Geological Survey Hydrologic Atlas HA-703, scale 1:500,000.

Liszewski, M.J., and Mann, L.J., 1993, Concentrations of 23 trace elements in ground water and surface water at and near the Idaho National Engineering Laboratory, Idaho, 1988–91: U.S. Geological Survey Open-File Report 93–126, DOE/ID-22110, 44 p. (Also available at http://pubs.er.usgs.gov/usgspubs/ofr/ofr93126.)

Olmsted, F.H., 1962, Chemical and physical character of ground water in the National Reactor Testing Station, Idaho: U.S. Atomic Energy Commission, Idaho Operations Office Publication IDO-22043-USGS, 142 p.

Parkhurst, D.L., Thorstenson, D.C., and Plummer, L.N., 1980, PHREEQE—A computer program for geochemical calculations: U.S. Geological Survey Water-Resources Investigations Report 80–96, 195 p. (Also available at http://pubs.er.usgs.gov/publication/wri8096.)

Plummer, L.N., Prestemon, E.C., and Parkhurst, D.L., 1991, An interactive code (NETPATH) for modeling NET geochemical reactions along a flow PATH: U.S. Geological Survey Water-Resources Investigations Report 91–4078, 130 p. (Also available at http://pubs.er.usgs.gov/publication/wri914078.)

Plummer, L.N., Rupert, M.G., Busenberg, E., and Schlosser, P., 2000, Age of irrigation water in groundwater from the Snake River Plain aquifer, South-Central Idaho: Ground Water, v. 38, p. 264–283.

Pollock, D.W., 1994, User's guide for MODPATH/ MODPATH-PLOT, version 3; a particle tracking post-processing package for MODFLOW, the U.S. Geological Survey finite-difference ground-water flow model: U.S. Geological Survey Open-File Report 94–464, 245 p. (Also available at http://pubs.er.usgs.gov/publication/ofr94464.)

Robertson, J.B., Schoen, Robert, and Barraclough, J.T., 1974, The influence of liquid waste disposal on the geochemistry of water at the National Reactor Testing Station, Idaho: 1952–1970: U.S. Geological Survey Open-File Report 73–238 (IDO-22053), 231 p. (Also available at http://pubs.er.usgs.gov/usgspubs/ofr/ofr73238.)

Schramke, J.A., Murphy, E.M., and Wood, B.D., 1996, The use of geochemical mass-balance and mixing models to determine groundwater sources: Applied Geochemistry, v. 1, no. 4, p. 523–539.

Spinazola, J.M., Tungate, A.M., and Rogers, T.L., 1992, Geohydrologic and chemical data from wells in the Mud Lake area, eastern Idaho, 1988–91: U.S. Geological Survey Open-File Report 92–133, 92 p. (Also available at http://pubs.er.usgs.gov/publication/ofr92133.)

Swanson, S.A., Rosentreter, J.J., Bartholomay, R.C., and Knobel, L.L., 2002, Geochemistry of the Little Lost River drainage basin, Idaho: U.S. Geological Survey Water-Resources Investigations Report 2002–4120, DOE/ID 22179, 29 p. (Also available at http://pubs.er.usgs.gov/usgspubs/wri/wri024120.)

Swanson, S.S., Rosentreter, J.J., Bartholomay, R.C., and Knobel, L.L., 2003, Geochemistry of the Birch Creek drainage basin, Idaho: U.S. Geological Survey Water-Resources Investigations Report 03–4272, DOE/ID-22188, 36 p. (Also available at http://pubs.er.usgs.gov/usgspubs/wri/wri034272.)

U.S. Geological Survey, 2011, National Water Information Systems-Web interface, USGS water data for Idaho: U.S. Geological Survey database, accessed in 2011 at http://waterdata.usgs.gov/id/nwis/nwis.

Warde, J.M., 1972, Lithium—Economic deposits, in Fairbridge, R.W., ed., The encyclopedia of geochemistry and environmental science: Stroudsburg, Pa., Dowden, Hutchinsen, and Ross, p. 662–664.

Appendix A. Modifications to the MODPATH Package

Weak Sinks and Sources

The MODPATH package is a particle tracking post-processing program developed to compute three-dimensional flow paths using output from groundwater-flow simulations by MODFLOW, the U.S. Geological Survey three-dimensional groundwater flow model (Pollock, 1994). In MODPATH, the user has the option of stopping a forward-moving particle when it enters a cell containing an internally distributed sink. Weak-sink cells are used to describe cases where some of the water flowing into a cell discharges to the sink and some passes through the cell. The user must decide whether particles entering these weak-sink cells discharge to the sink or pass through the cell. MODPATH accounts for weak sinks by stopping particles where the discharge to sinks is larger than a specified fraction of the total inflow to the cell, or

$$F = \frac{-QSS}{QI} > FRAC \tag{1}$$

where F is the fraction of the total inflow to the cell; QSS is the discharge to the sink; QI is the total inflow to the cell; and FRAC is the user defined fraction of the total inflow to the cell.

The equivalent of a "weak sink" in backward-particle tracking is a "weak source", where groundwater velocities in a backward-particle tracking simulation are multiplied by -1 and the flow direction is reversed. A weak-source cell describes the case where some water flowing out of a cell originates from an internal source and some water passes through the cell from adjacent cells. Particles entering a weak-source cell should terminate at the source or pass though the cell. In the current version of MODPATH (version 5.0) weak sources are unaccounted for and it is a mistake to assume that weak sinks are treated as weak sources during backward-particle tracking. MODPATH source code was modified to account for weak sources in backward-particle tracking.

Recompiling the Source Code

Any modification to the MODPATH package requires a recompilation of its source code. Version 5 of the source code is provided in the MODPATH distribution file (*mpath5_0.exe*) and is available for download at http://water.usgs.gov/nrp/gwsoftware/modpath5/modpath5.html. The original source code was written using options specific to the Lahey/Fujitsu Fortran 95 compiler (version 5.7); therefore, code alterations were necessary to compile using gfortran, a Fortran 95/2003 compiler that is part of the GNU Compiler Collection (http://gcc.gnu.org/) (version 4.5.0) and available under the GNU General Public License. Recompiling with gfortran required the following changes:

1. After line 145 of the *Budget.for* file, added a statement to recall the last value assigned to the TXTSAV parameter; that is

        IF(IOLD.EQ.1) TXTSAV='   STORAGE'

2. On lines 116 and 167 of the *Flowdata.for* file, changed the FORM='BINARY' declaration in the OPEN statement to FORM='UNFORMATTED'.

3. On line 79 of the *MPATH5.FOR* file, removed OPEN statement.

4. On line 17 of the *openspec.inc* file, changed the DATA ACCESS statement from /'SEQUENTIAL'/ to /'STREAM'/.

5. On line 29 of the *openspec.inc* file, uncommented DATA FORM/'UNFORMATTED'/ statement.

6. On line 35 of the *openspec.inc* file, removed DATA FORM/'BINARY'/ statement.

7. On lines 228 and 303 of the *utilmp.for* file, removed the UNFORMATTED=FM declaration in the INQUIRE statement.

Adding the Weak-Source Feature

The addition of a weak-source feature to the MODPATH package required only minor changes to the original source code; these changes are as follows:

1. On lines 300 and 301, of the *Mpmove.for* file, removed the `IF` statement. The removed lines of code are

    ```
    IF(IBOUND(JP,IP,KP).GT.-1000.AND.IBOUND(JP,IP,KP).LT.1000)
    1  GO TO 80
    ```

    where `IBOUND` is the boundary array containing cell types. An `IBOUND` element less than 0 is a specified hydraulic head cell, equal to zero is an inactive cell, and greater than 0 is an active cell. The user specified `IBOUND` elements are required to be greater than -1,000 and less than 1,000. In *Flowdata.for*, elements in the `IBOUND` array corresponding to cells with internal sinks (or sources) are multiplied by 1,000; therefore, only those cells containing internal sinks are considered in the weak-sink evaluation (eq. 1). Note that `IBOUND` values corresponding to cells assigned with a directional component of flow to any of the six cell faces using `IFACE` are not multiplied by 1,000. Removal of the `IBOUND` conditional statement from *Mpmove.for* permits all cells to be checked to determine whether they meet the criterion of a weak-sink or weak-source.

2. On line 345 of the *Mpmove.for* file, accounted for backward particle tracking by multiplying `QSS`, the internal source (sink) flow rate, by `VSIGN`, where `VSIGN` is equal to 1 for forward particle tracking and -1 for backward particle tracking. The revised statement is expressed as

    ```
    IF((VSIGN*QSS(JP,IP,KP)).GE.0.0) GO TO 80
    ```

3. On line 354 of the *Mpmove.for* file, accounted for backward particle tracking by multiplying `QSS` by `VSIGN`. The revised statement is expressed as,

    ```
    F= -1.0*(VSIGN*QSS(JP,IP,KP))/QI
    ```

    where `F` is the fraction of the total inflow to the cell (eq. 1).

Reference Cited

Pollock, D.W., 1994, User's guide for MODPATH/MODPATH-PLOT, version 3; a particle tracking post-processing package for MODFLOW, the U.S. Geological Survey finite-difference ground-water flow model: U.S. Geological Survey Open-File Report 94–464, 245 p. (Also available at http://pubs.er.usgs.gov/publication/ofr94464.)

Appendix B. Data for Surface-Water and Groundwater Sampling Sites Located at the Idaho National Laboratory and Vicinity, Idaho

[**Local name:** is the local well identifier used in this study. **Longitude and Latitude:** in degrees, minutes, seconds and based on NAD 27 datum. **Land-surface altitude:** in feet above mean sea level and based on NGVD 29 datum (ft amsl). **Site type:** Well, monitoring well; MLMS, multi-level monitoring systems; Surface, surface water; Spring, thermal springs; Pond, waste-disposal ponds. **Well and hole depth:** in feet below land surface. **Site identifier:** is the unique numerical identifiers used to access well data (http://waterdata.usgs.gov/nwis). **Open interval:** for multi-level systems the open interval is a hydraulically isolated depth interval. **Model layer:** See Ackerman and others (2010) for more information. Layer 1 represents water from the upper 100 feet of the saturated aquifer; layer 2 represents water from 100 to 200 feet below the top of the saturated aquifer; layer 3 represents water from 200 to 300 feet below the top of the saturated aquifer; layer 4 represents water from 300 to 500 feet below the top of the saturated aquifer; layer 5 represents water from 500 to 800 feet below the top of the saturated aquifer; and layer 6 represents water from greater than 800 feet below the top of the saturated aquifer. **Abbreviations:** ft amsl, feet above mean sea level; ft bls, feet below land surface; NA, not applicable; OMD, outside model domain; –, not available]

Local name	Longitude	Latitude	Land-surface altitude (ft amsl)	Site type	Well depth (ft bls)	Hole depth (ft bls)	Site identifier	Open interval (ft bls)	Model layer
ANP 6	112°44'31.39"	43°51'51.57"	4,794.43	Well	305	305	435152112443101	211–256	1
								266–296	1
ANP 9	112°40'00.36"	43°48'55.71"	4,786.14	Well	322	322	434856112400001	237–314	1
Arbor Test 1	112°38'48.01"	43°35'08.92"	5,163.95	Well	790	790	433509112384801	680–731	1
Arco City Well 4	113°18'17"	43°37'58"	5,320.00	Well	250	250	433758113181701	209–217	OMD
								225–242	OMD
Area 2	112°47'02.40"	43°32'22.62"	5,128.60	Well	877	877	433223112470201	–	–
Atomic City	112°48'41"	43°26'38"	5,017.00	Well	–	–	432638112484101	35–639	1
BFW	112°53'51"	43°30'42"	–	Well	–	–	433042112535101	–	–
CFA 1	112°56'19.53"	43°32'03.83"	4,927.98	Well	639	685	433204112562001	444–639	1, 2
CFA 2	112°56'35.11"	43°31'43.94"	4,931.70	Well	681	681	433144112563501	521–651	1, 2
								661–681	2
EBR 1	113°00'26.37"	43°30'49.65"	5,024.32	Well	1,075	1,075	433051113002601	600–750	1, 2
								750–1,075	2–4
Engberson Well (ML-9)	112°26'45"	43°50'28"	–	Well	281	281	435028112264501	109–281	OMD
Fire Station 2	112°56'23"	43°35'48"	4,902.31	Well	510	518	433548112562301	427–467	1
								501–511	1
IET 1 Disposal	112°42'05.22"	43°51'53.38"	4,790.02	Well	242	329	435153112420501	219–319	1, 2
INEL-1 WS	112°56'36.17"	43°37'17.06"	4,873.29	Well	10,333	10,365	433717112563501	4,210–4,225	OMD
								4,240–4,270	OMD
								4,300–4,315	OMD
								4,490–4,520	OMD
								4,775–4,790	OMD
								5,085–5,100	OMD
								5,230–5,245	OMD
								5,995–6,010	OMD
								6,220–6,235	OMD
								6,260–6,275	OMD
								6,796–10,333	OMD
								10,333–10,365	OMD
Leo Rogers 1	112°50'49"	43°25'33"	5,039.00	Well	720	720	432533112504901	20–720	1, 2
Neville Well (ML-7)	112°09'29"	43°55'40"	4,830.00	Well	85	85	435540112092901	–	OMD
NPR Test	112°52'31.26"	43°34'49.43"	4,933.13	Well	600	600	433449112523101	500–535	1
NPR-W01	112°52'31.07"	43°34'50.80"	4,929.92	Well	520	527	–	–	1

Appendix B.   Data for surface-water and groundwater sampling sites located at the Idaho National Laboratory and vicinity, Idaho.—Continued

[**Local name:** is the local well identifier used in this study. **Longitude and Latitude:** in degrees, minutes, seconds and based on NAD 27 datum. **Land-surface altitude:** in feet above mean sea level and based on NGVD 29 datum (ft amsl). **Site type:** Well, monitoring well; MLMS, multi-level monitoring systems; Surface, surface water; Spring, thermal springs; Pond, waste-disposal ponds. **Well and hole depth:** in feet below land surface. **Site identifier:** is the unique numerical identifiers used to access well data (http://waterdata.usgs.gov/nwis). **Open interval:** for multi-level systems the open interval is a hydraulically isolated depth interval. **Model layer:** See Ackerman and others (2010) for more information. Layer 1 represents water from the upper 100 feet of the saturated aquifer; layer 2 represents water from 100 to 200 feet below the top of the saturated aquifer; layer 3 represents water from 200 to 300 feet below the top of the saturated aquifer; layer 4 represents water from 300 to 500 feet below the top of the saturated aquifer; layer 5 represents water from 500 to 800 feet below the top of the saturated aquifer; and layer 6 represents water from greater than 800 feet below the top of the saturated aquifer. **Abbreviations:** ft amsl, feet above mean sea level; ft bls, feet below land surface; NA, not applicable; OMD, outside model domain; –, not available]

Local name	Longitude	Latitude	Land-surface altitude (ft amsl)	Site type	Well depth (ft bls)	Hole depth (ft bls)	Site identifier	Open interval (ft bls)	Model layer
Pancheri 6	113°10'40.4"	43°57'28.1"	5,375.00	Well	87	87	435728113103701	37–87	OMD
Park Bell (08N 34E 17CCC7)	112°29'36"	44°00'58"	4,808.92	Well	48	50	440058112293605	–	OMD
PSTF Test	112°45'41.46"	43°49'40.74"	4,786.35	Well	319	322	434941112454201	190–316	1
P&W 2	112°45'30.80"	43°54'19.00"	4,890.86	Well	386	386	435419112453101	313–383	1
RWMC M3S	113°02'18.31"	43°30'08.29"	5,016.34	Well	633	660	433008113021801	603–633	1
RWMC M7S	113°01'48.39"	43°30'22.56"	5,005.15	Well	628	638	433023113014801	598–628	1
Site 4	112°54'20"	43°36'18"	4,795.03	Well	495	495	433617112542001	416–491	1, 2
Site 9	112°53'00.80"	43°31'22.86"	4,925.65	Well	1,057	1,131	433123112530101	681–1,057	3–5
Site 14	112°46'31.50"	43°43'34.66"	4,793.52	Well	717	717	434334112463101	535–716	3, 4
Site 17	112°57'56.50"	43°40'26.74"	4,880.47	Well	600	600	434027112575701	15–600	1–3
Site 19	112°58'21.49"	43°35'22.32"	4,925.95	Well	860	865	433522112582101	472–512	1
								533–572	1, 2
								597–617	2
								781–863	OMD
TAN Exploration	112°45'32.76"	43°50'38.79"	4,784.30	Well	–	–	435038112453401	267–550	1–4
USGS 1	112°47'08.54"	43°27'00.08"	5,022.34	Well	636	636	432700112470801	600–630	1
USGS 2	112°43'21.28"	43°33'19.87"	5,125.22	Well	699	704	433320112432301	675–696	1
USGS 4	112°28'21.62"	43°46'55.93"	4,790.73	Well	553	553	434657112282201	285–315	1
								322–553	1–3
USGS 5	112°49'37.65"	43°35'42.75"	4,937.57	Well	494	500	433543112493801	475–497	1
USGS 6	112°45'36.66"	43°40'31.12"	4,898.55	Well	620	620	434031112453701	452–475	1
								532–620	2, 3
USGS 7	112°44'39.87"	43°49'14.81"	4,789.24	Well	903	1,200	434915112443901	239–252	1
								252–775	1–5
								241–261	1
								241–261	1
								760–940	5
								940–1,200	5, 6
USGS 8	113°11'57.43"	43°31'20.51"	5,194.94	Well	812	812	433121113115801	782–812	1
USGS 9	113°04'39.78"	43°27'32.38"	5,030.32	Well	654	654	432740113044501	618–648	1
								652–654	1
USGS 11	113°06'42.52"	43°23'36.18"	5,067.12	Well	704	704	432336113064201	673–704	1
USGS 12	112°55'07.10"	43°41'26.19"	4,819.00	Well	563	692	434126112550701	587–692	3, 4

Appendix B. Data for surface-water and groundwater sampling sites located at the Idaho National Laboratory and vicinity, Idaho.—
Continued

[**Local name:** is the local well identifier used in this study. **Longitude and Latitude:** in degrees, minutes, seconds and based on NAD 27 datum. **Land-surface altitude:** in feet above mean sea level and based on NGVD 29 datum (ft amsl). **Site type:** Well, monitoring well; MLMS, multi-level monitoring systems; Surface, surface water; Spring, thermal springs; Pond, waste-disposal ponds. **Well and hole depth:** in feet below land surface. **Site identifier:** is the unique numerical identifiers used to access well data (http://waterdata.usgs.gov/nwis). **Open interval:** for multi-level systems the open interval is a hydraulically isolated depth interval. **Model layer:** See Ackerman and others (2010) for more information. Layer 1 represents water from the upper 100 feet of the saturated aquifer; layer 2 represents water from 100 to 200 feet below the top of the saturated aquifer; layer 3 represents water from 200 to 300 feet below the top of the saturated aquifer; layer 4 represents water from 300 to 500 feet below the top of the saturated aquifer; layer 5 represents water from 500 to 800 feet below the top of the saturated aquifer; and layer 6 represents water from greater than 800 feet below the top of the saturated aquifer. **Abbreviations:** ft amsl, feet above mean sea level; ft bls, feet below land surface; NA, not applicable; OMD, outside model domain; –, not available]

Local name	Longitude	Latitude	Land-surface altitude (ft amsl)	Site type	Well depth (ft bls)	Hole depth (ft bls)	Site identifier	Open interval (ft bls)	Model layer
USGS 14	112°56'31.92"	43°20'19.27"	5,132.88	Well	752	752	432019112563201	720–747	1
USGS 15	112°55'17.35"	43°42'34.84"	4,811.99	Well	610	1,497	434234112551701	540–610	3
								610–1,497	3, 4
USGS 17	112°51'54.27"	43°39'36.42"	4,833.44	Well	498	498	433937112515401	438–445	1
								496–498	2
USGS 18	112°44'09.29"	43°45'40.70"	4,804.23	Well	329	329	434540112440901	298–322	1
USGS 19	112°57'56.58"	43°44'26.68"	4,800.06	Well	399	405	434426112575701	285–306	1
USGS 22	113°03'21.09"	43°34'22.28"	5,048.27	Well	657	657	433422113031701	619–634	1
								644–657	1
USGS 23	113°00'00.02"	43°40'55.15"	4,884.20	Well	458	467	434055112595901	410–430	1
USGS 26	112°39'40.74"	43°52'10.55"	4,788.69	Well	267	267	435212112394001	232–267	1
USGS 27	112°32'18.90"	43°48'51.22"	4,783.90	Well	312	312	434851112321801	250–260	1
								298–308	1
USGS 29	112°28'50.25"	43°44'06.86"	4,877.48	Well	426	426	434407112285101	363–398	1
								398–426	1
USGS 31	112°34'20.47"	43°46'25.90"	4,785.79	Well	428	428	434625112342101	285–305	1
								306–428	1, 2
USGS 32	112°32'21.24"	43°44'44.07"	4,812.02	Well	392	392	434444112322101	306–324	1
								324–392	1, 2
USGS 36	112°56'51.47"	43°33'30.11"	4,928.83	Well	567	567	433330112565201	430–567	1
USGS 82	112°55'10.34"	43°34'00.93"	4,906.83	Well	693	700	433401112551001	470–570	1, 2
								593–700	2, 3
USGS 83	112°56'15.28"	43°30'23.03"	4,941.11	Well	752	752	433023112561501	516–752	1–3
USGS 86	113°08'01.44"	43°29'34.79"	5,076.92	Well	691	691	432935113080001	48–691	1
USGS 89	113°03'31.73"	43°30'05.67"	5,030.24	Well	637	650	433005113032801	576–650	1
USGS 97	112°55'16.76"	43°38'06.77"	4,858.49	Well	510	510	433807112551501	388–510	1, 2
USGS 98	112°56'36"	43°36'57"	4,882.64	Well	508	508	433657112563601	407–505	1
								401–421	OMD
								463–505	1
								418–428	1
								468–508	1
								401–421	OMD
								463–505	1
USGS 99	112°55'21.16"	43°37'03.74"	4,871.55	Well	440	450	433705112552101	303–449	1
								449–450	1
USGS 100	112°40'06.67"	43°35'02.72"	5,157.94	Well	750	750	433503112400701	662–750	1

Appendix B. Data for surface-water and groundwater sampling sites located at the Idaho National Laboratory and vicinity, Idaho.—Continued

[**Local name:** is the local well identifier used in this study. **Longitude and Latitude:** in degrees, minutes, seconds and based on NAD 27 datum. **Land-surface altitude:** in feet above mean sea level and based on NGVD 29 datum (ft amsl). **Site type:** Well, monitoring well; MLMS, multi-level monitoring systems; Surface, surface water; Spring, thermal springs; Pond, waste-disposal ponds. **Well and hole depth:** in feet below land surface. **Site identifier:** is the unique numerical identifiers used to access well data (http://waterdata.usgs.gov/nwis). **Open interval:** for multi-level systems the open interval is a hydraulically isolated depth interval. **Model layer:** See Ackerman and others (2010) for more information. Layer 1 represents water from the upper 100 feet of the saturated aquifer; layer 2 represents water from 100 to 200 feet below the top of the saturated aquifer; layer 3 represents water from 200 to 300 feet below the top of the saturated aquifer; layer 4 represents water from 300 to 500 feet below the top of the saturated aquifer; layer 5 represents water from 500 to 800 feet below the top of the saturated aquifer; and layer 6 represents water from greater than 800 feet below the top of the saturated aquifer. **Abbreviations:** ft amsl, feet above mean sea level; ft bls, feet below land surface; NA, not applicable; OMD, outside model domain; –, not available]

Local name	Longitude	Latitude	Land-surface altitude (ft amsl)	Site type	Well depth (ft bls)	Hole depth (ft bls)	Site identifier	Open interval (ft bls)	Model layer
USGS 101	112°38'19.91"	43°32'55.75"	5,251.16	Well	842	865	433255112381801	750–865	1
USGS 102	112°55'16.43"	43°38'50.87"	4,850.28	Well	445	445	433853112551601	359–445	1
USGS 103	112°56'06.53"	43°27'13.57"	5,007.42	Well	760	760	432714112560701	575–760	1, 2
				MLMS	1,297	1,307	432714112560723	669.6–691.3	2
							432714112560720	766.9–832.1	3
							432714112560716	891.6–919.6	4
							432714112560712	958.0–1,013.5	4
							432714112560708	1,063.2–1,097.6	4–5
							432714112560704	1,184.4–1,239.9	5
							432714112560702	1,257.4–1,279.4	5
USGS 104	112°56'08.14"	43°28'56.07"	4,987.64	Well	700	700	432856112560801	550–700	1, 2
USGS 105	113°00'17.78"	43°27'03.40"	5,095.12	Well	1,300	1,409	432703113001801	400–800	1, 2
				MLMS	1,300	1,409	432703113001818	706.9–751.9	1
							432703113001815	830.4–862.3	2, 3
							432703113001811	929.3–982.4	3, 4
							432703113001807	1,034.6–1,102.4	4
							432703113001803	1,224.8–1,276.2	5
USGS 107	112°53'27.53"	43°29'42.02"	4,917.50	Well	690	690	432942112532801	270–690	1–3
USGS 108	112°58'26.34"	43°26'58.79"	5,031.36	Well	1,218	1,218	432659112582601	400–760	1, 2
								755–765	2
				MLMS	1,218	1,218	432659112582616	642.1–678.8	1
							432659112582615	681.8–788.4	1, 2
							432659112582613	791.4–829.7	3
							432659112582612	832.7–869.0	3
							432659112582610	872.0–903.7	3, 4
							432659112582609	906.7–977.4	4
							432659112582608	980.4–1,015.0	4
							432659112582606	1,018.0–1,059.6	4
							432659112582605	1,062.6–1,118.6	4, 5
							432659112582604	1,121.6–1,157.9	5
							432659112582602	1,160.9–1,191.9	5
USGS 109	113°02'55.81"	43°27'01.23"	5,043.64	Well	800	800	432701113025601	348–800	1, 2
								615–795	1, 2
								795–800	2
USGS 110A	112°50'14.72"	43°27'17.13"	4,999.46	Well	644	657	432717112501502	240–657	1, 2
USGS 112	112°56'30.74"	43°33'14.50"	4,927.82	Well	507	563	433314112563001	432–444	OMD
								444–563	1
USGS 113	112°56'18.29"	43°33'14.53"	4,925.32	Well	556	564	433314112561801	445–564	1
USGS 115	112°55'41.39"	43°33'20.22"	4,918.86	Well	581	581	433320112554101	437–581	1, 2

Appendix B.   Data for surface-water and groundwater sampling sites located at the Idaho National Laboratory and vicinity, Idaho.—Continued

[**Local name:** is the local well identifier used in this study. **Longitude and Latitude:** in degrees, minutes, seconds and based on NAD 27 datum. **Land-surface altitude:** in feet above mean sea level and based on NGVD 29 datum (ft amsl). **Site type:** Well, monitoring well; MLMS, multi-level monitoring systems; Surface, surface water; Spring, thermal springs; Pond, waste-disposal ponds. **Well and hole depth:** in feet below land surface. **Site identifier:** is the unique numerical identifiers used to access well data (http://waterdata.usgs.gov/nwis). **Open interval:** for multi-level systems the open interval is a hydraulically isolated depth interval. **Model layer:** See Ackerman and others (2010) for more information. Layer 1 represents water from the upper 100 feet of the saturated aquifer; layer 2 represents water from 100 to 200 feet below the top of the saturated aquifer; layer 3 represents water from 200 to 300 feet below the top of the saturated aquifer; layer 4 represents water from 300 to 500 feet below the top of the saturated aquifer; layer 5 represents water from 500 to 800 feet below the top of the saturated aquifer; and layer 6 represents water from greater than 800 feet below the top of the saturated aquifer. **Abbreviations:** ft amsl, feet above mean sea level; ft bls, feet below land surface; NA, not applicable; OMD, outside model domain; –, not available]

Local name	Longitude	Latitude	Land-surface altitude (ft amsl)	Site type	Well depth (ft bls)	Hole depth (ft bls)	Site identifier	Open interval (ft bls)	Model layer
USGS 116	112°55'32.67"	43°33'31.55"	4,916.05	Well	572	580	433331112553201	400–438 438–572	OMD 1, 2
USGS 117	113°02'58.67"	43°29'54.50"	5,012.50	Well	655	655	432955113025901	550–655	1
USGS 120	113°03'14.01"	43°29'19.19"	5,040.43	Well	705	705	432919113031501	638–705	1
USGS 124	112°58'28.22"	43°23'06.85"	5,102.30	Well	800	800	432307112583101	750–800	1, 2
USGS 125	113°05'30.37"	43°25'59.41"	5,050.71	Well	774	774	432602113052801	620–774	1, 2
Wagoner Ranch	112°53'22"	44°08'13"	6,010.00	Well	295	295	440813112532201	294–295	OMD
BLR Mackay Dam	113°38'50"	43°56'21"	5,946.39	Surface	NA	NA	13127000	NA	NA
BLR Mackay Brdg	113°34'39"	43°53'14"	–	Surface	NA	NA	13127780	NA	NA
BLR Lincoln Blvd	112°56'33"	43°34'26"	4,900.00	Surface	NA	NA	13132535	NA	NA
Birch Crk at Blue Dome	112°54'30.67"	44°09'12.44"	6,050.00	Surface	NA	NA	13117020	NA	NA
Camas Crk Mud Lake	112°21'26"	43°53'29"	4,774.99	Surface	NA	NA	13115000	NA	NA
Lidy Hot Springs	112°33'10"	44°08'32"	5,260.00	Spring	NA	NA	440832112331001	NA	NA
LLR near INEEL	112°59'49.21"	43°47'02.15"	–	Surface	NA	NA	–	NA	NA
LLR north of Howe	113°06'00"	43°53'10"	5,020.00	Surface	NA	NA	13119000	NA	NA
Stoddart	112°33'21"	43°54'02"	4,784.00	Well	207	207	435402112332101	–	OMD
Reno Ranch	112°42'55"	44°01'42"	5,110.00	Well	540	540	440142112425501	–	–
USGS 126 A	112°47'12.90"	43°55'28.75"	4,988.69	Well	648	648	435529112471301	–	–
USGS 126 B	112°47'13.67"	43°55'28.51"	4,989.25	Well	472	472	435529112471401	–	–
6N-34E-32ACD1	112°28'48"	43°48'18"	4,788.00	Well	NA	NA	434818112284801	–	–
6N-35E-12BCD1	112°17'27"	43°51'50"	4,790.00	Well	150	150	435150112172701	9–150	OMD
6N 35E 21AAB1	112°20'26"	43°50'28"	4,784.50	Well	276	276	435028112202601	–	OMD
7N-33E-16BAB1	112°35'17"	43°56'32"	4,790.00	Well	340	340	435632112351701	140–– 244–340	OMD OMD
7N-34E-10ACA1	112°26'33.3"	43°57'13.3"	4,805.00	Well	88	88	435712112263201	–	OMD
7N-35E-22DAD1	112°19'02"	43°55'05"	4,792.49	Well	43	43	435505112190201	–	OMD
7N-36E-5CAA1	112°14'50.4"	43°57'52.3"	4,798.00	Well	239	239	435753112145101	75–239	OMD
8N-37E-27BBC1	112°05'45"	43°59'49"	4,975.00	Well	–	–	435949112054501	–	OMD
8N-37E-30ABC1	112°08'47"	43°59'51"	4,890.00	Well	–	–	435951112084701	–	OMD
Grazing Well 2	112°49'20"	43°15'53"	4,771.67	Well	390	–	431553112492001	–	OMD
Grazing Service CCC 3	112°58'54"	43°09'11"	4,849.00	Well	451	451	430911112585401	437–451	OMD

Appendix B.   Data for surface-water and groundwater sampling sites located at the Idaho National Laboratory and vicinity, Idaho.—Continued

[**Local name:** is the local well identifier used in this study. **Longitude and Latitude:** in degrees, minutes, seconds and based on NAD 27 datum. **Land-surface altitude:** in feet above mean sea level and based on NGVD 29 datum (ft amsl). **Site type:** Well, monitoring well; MLMS, multi-level monitoring systems; Surface, surface water; Spring, thermal springs; Pond, waste-disposal ponds. **Well and hole depth:** in feet below land surface. **Site identifier:** is the unique numerical identifiers used to access well data (http://waterdata.usgs.gov/nwis). **Open interval:** for multi-level systems the open interval is a hydraulically isolated depth interval. **Model layer:** See Ackerman and others (2010) for more information. Layer 1 represents water from the upper 100 feet of the saturated aquifer; layer 2 represents water from 100 to 200 feet below the top of the saturated aquifer; layer 3 represents water from 200 to 300 feet below the top of the saturated aquifer; layer 4 represents water from 300 to 500 feet below the top of the saturated aquifer; layer 5 represents water from 500 to 800 feet below the top of the saturated aquifer; and layer 6 represents water from greater than 800 feet below the top of the saturated aquifer. **Abbreviations:** ft amsl, feet above mean sea level; ft bls, feet below land surface; NA, not applicable; OMD, outside model domain; –, not available]

Local name	Longitude	Latitude	Land-surface altitude (ft amsl)	Site type	Well depth (ft bls)	Hole depth (ft bls)	Site identifier	Open interval (ft bls)	Model layer
Houghland Well	113°07'14.7"	43°14'39.3"	5,110.00	Well	775	775	431439113071401	6–475	OMD
								475–775	OMD
Crossroads Well	113°09'27"	43°21'28"	5,120.00	Well	796	796	432128113092701	17–748	1
								774–796	1
Fingers Butte Well	113°16'53"	43°24'24"	5,364.00	Well	1,056	1,056	432424113165301	994–1,056	1
USGS 20	112°54'59.41"	43°32'52.79"	4,915.11	Well	676	676	433253112545901	417–477	1
								515–553	1
USGS 57	112°56'26.00"	43°33'44.04"	4,922.23	Well	582	732	433344112562601	477–732	1–3
USGS 65	112°57'47.13"	43°34'46.85"	4,924.75	Well	498	498	433447112574501	456–472	1
								472–498	1
USGS 85	112°57'11.89"	43°32'46.23"	4,938.99	Well	614	637	433246112571201	522–614	1, 2
USGS 88	113°03'01.96"	43°29'40.20"	5,020.81	Well	663	663	432940113030201	587–635	1
USGS 119	113°02'33.69"	43°29'44.61"	5,031.84	Well	705	705	432945113023401	639–705	1, 2
USGS 121	112°56'03.32"	43°34'49.48"	4,909.66	Well	475	746	433450112560301	449–475	1
USGS 122	112°55'51.41"	43°33'53.45"	4,913.79	Well	480	483	433353112555201	449–475	1
USGS 123	112°56'13.73"	43°33'51.70"	4,919.26	Well	515	744	433352112561401	450–475	1
CPP Pond 1	112°55'51"	43°33'51"	4,918.00	Pond	NA	NA	433351112555101	NA	NA
McKinney (10N 29E 24AAD1)	112°56'06"	44°11'13"	6,205.00	Well	43	43	441113112560601	–	OMD
No Name 1	112°45'32.76"	43°50'38.79"	4,784.30	Well	–	–	435038112453401	267–550	1–4
Simplot 1 (05N 29E 01BBB1) (Ruby Farms Well)	112°57'18"	43°47'51"	4,805.00	Well	149	154	434751112571801	106–149	OMD
ANP 10	112°40'03.89"	43°49'09.07"	4,786.05	Well	676	681	434909112400401	552–677	4
EOCR 1	112°53'51.10"	43°31'20.14"	4,939.00	Well	1,237	1,237	433120112535101	1,051–1,237	5
FET 1	112°43'20"	43°51'20"	4,780.76	Well	330	339	435120112432101	230–330	1, 2
FET 2	112°43'17.90"	43°51'19.09"	4,780.90	Well	455	462	435119112431801	209–448	1–3
GCRE 1	112°49'41"	43°31'50"	5,059.00	Well	–	–	–	–	–
TRA 1	112°57'37.68"	43°35'21.46"	4,917.35	Well	600	600	433521112573801	481–581	1, 2
TRA 2	112°57'48.91"	43°35'21.76"	4,919.83	Well	747	772	433523112575001	558–567	1, 2
								572–601	2
OMRE	112°53'46.28"	43°31'16.68"	4,937.48	Well	743	943	433116112534701	535–626	1, 2
								920–938	4

Appendix B.   Data for surface-water and groundwater sampling sites located at the Idaho National Laboratory and vicinity, Idaho.—Continued

[**Local name:** is the local well identifier used in this study. **Longitude and Latitude:** in degrees, minutes, seconds and based on NAD 27 datum. **Land-surface altitude:** in feet above mean sea level and based on NGVD 29 datum (ft amsl). **Site type:** Well, monitoring well; MLMS, multi-level monitoring systems; Surface, surface water; Spring, thermal springs; Pond, waste-disposal ponds. **Well and hole depth:** in feet below land surface. **Site identifier:** is the unique numerical identifiers used to access well data (http://waterdata.usgs.gov/nwis). **Open interval:** for multi-level systems the open interval is a hydraulically isolated depth interval. **Model layer:** See Ackerman and others (2010) for more information. Layer 1 represents water from the upper 100 feet of the saturated aquifer; layer 2 represents water from 100 to 200 feet below the top of the saturated aquifer; layer 3 represents water from 200 to 300 feet below the top of the saturated aquifer; layer 4 represents water from 300 to 500 feet below the top of the saturated aquifer; layer 5 represents water from 500 to 800 feet below the top of the saturated aquifer; and layer 6 represents water from greater than 800 feet below the top of the saturated aquifer. **Abbreviations:** ft amsl, feet above mean sea level; ft bls, feet below land surface; NA, not applicable; OMD, outside model domain; –, not available]

Local name	Longitude	Latitude	Land-surface altitude (ft amsl)	Site type	Well depth (ft bls)	Hole depth (ft bls)	Site identifier	Open interval (ft bls)	Model layer
SL-1 1	112°49'23"	43°31'11"	5,053.00	Well	–	–	–	–	–
SPERT 1	112°52'03"	43°32'53"	4,924.98	Well	653	653	433252112520301	482–492	1
								522–542	1
								552–582	1, 2
								597–617	2
								632–652	2
SPERT 2	112°51'51"	43°32'46"	4,924.10	Well	1,217	1,217	433247112515201	951–1,217	4, 5
USGS 30	112°31'54.36"	43°46'00.72"	4,793.87	Well	300	1,007	434601112315401	290–300	1
USGS 132	113°02'50.93"	43°29'06.68"	5,028.60	MLMS	1,238	1,238	432906113025022	623.6–659.6	1
							432906113025018	726.6–787.1	2
							432906113025014	811.5–863.8	3
							432906113025010	911.1–935.4	4
							432906113025006	984.3–1,043.1	4
							432906113025001	1,152.3–1,213.6	5
USGS 133	112°55'43.80"	43°36'05.50"	4,890.12	MLMS	798	818	433605112554312	448.0–480.2	1
							433605112554308	555.5–590.7	2
							433605112554305	685.5–695.6	2, 3
							433605112554301	724.8–766.4	4
USGS 134	112°59'58.27"	43°36'11.15"	4,968.84	MLMS	894	949	433611112595819	553.8–589.8	1
							433611112595815	638.9–651.9	2
							433611112595811	690.9–720.0	2, 3
							433611112595807	782.0–818.0	3, 4
							433611112595804	846.0–868.0	4
							433611112595803	846.0–868.0	4
USGS 135	113°09'36"	43°27'53"	5,150.00	MLMS	1,157	1,198	432753113093613	726.9–762.2	1
							432753113093609	822.5–861.1	1, 2
							432753113093605	967.4–1,007.5	3
							432753113093601	1,105.5–1,140.0	4
MIDDLE 2050A	112°57'05.38"	43°34'09.48"	4,928.22	MLMS	1,376	1,427	433409112570515	464.9–538.6	1
							433409112570512	643.3–703.4	2, 3
							433409112570509	790.0–807.4	4
							433409112570506	998.7–1,040.6	OMD
							433409112570503	1,179.7–1,226.7	OMD
MIDDLE 2051	113°00'49.38"	43°32'16.93"	4,997.31	MLMS	1,175	1,179	433217113004912	561.8–609.2	1
							433217113004909	748.4–770.8	2, 3
							433217113004906	826.2–876.4	3, 4
							433217113004903	1,090.5–1,127.5	OMD
							433217113004901	1,140.3–1,176.5	OMD

Appendix C. Data for Streams Used in the Steady-State Model of Groundwater Flow, Idaho National Laboratory and Vicinity, Idaho

[**Map No.:** Identifier used to locate streamflow-gaging stations and stream reaches on the map in figure 3. **Site identifier:** Unique numerical identifier used to access streamflow-gaging station data (http://waterdata.usgs.gov/nwis). **Local name:** Local streamflow-gaging station or stream-reach identifier used in this study. **Abbreviation:** NA, not applicable]

Map No.	Site identifier	Local name
		Streamflow-gaging stations
500	13127000	Big Lost River below Mackay Reservoir near Mackay
501	12132500	Big Lost River near Arco
502	13132513	INL Diversion at Head, near Arco
503	13132515	INL Diversion at Outlet of Spreading Area A near Arco
504	13132520	Big Lost River below INL diversion, near Arco
505	13132535	Big Lost River at Lincoln Boulevard Bridge near Atomic City
506	13132565	Big Lost River above Big Lost River Sinks near Howe
		Stream reaches
600	NA	Big Lost River above Arco gaging station (13132500)
601	NA	Big Lost river between Arco gaging station (13132500) and INL Diversion
602	NA	Spreading Area A
603	NA	Spreading Area B
604	NA	Spreading Area C
605	NA	Spreading Area D
606	NA	Big Lost River between INL diversion and Lincoln Boulevard gaging station (13132535)
607	NA	Big Lost River between Lincoln Boulevard gaging station (13132535) and Big Lost River Sinks gaging station (13132565)
608	NA	Big Lost River Sinks
609	NA	Big Lost River Playas 1 and 2
610	NA	Big Lost River Playa 3
611	NA	Little Lost River
612	NA	Birch Creek power diversion return

Appendix D. Cation and Anion Data for Selected Wells at the Idaho National Laboratory, Idaho

[Local name: is local well identifier used in this study. Concentrations are in milligrams per liter. Site identifier: is the unique numerical identifiers used to access well data (http://waterdata.usgs.gov/nwis). Wells with compositional and temporal variations in water type as noted by Olmsted (1962) are indicated in bold. Water type: is based on criteria used by Olmsted (1962). Abbreviations: Ca, calcium; Mg, magnesium; Sr, strontium; Na, sodium; K, potassium; CO3, carbonate; HCO3, bicarbonate; SO4, sulfate; Cl, chloride; NO3, nitrate; F, fluoride; mEq, milliequivalents; %, percent; –, not sampled]

Local name	Site identifier	Ca	Mg	Ca+Mg (mEq)	Ca+Mg % (mEq)	Sr	Ca+Mg+Sr (mEq)	Na	K	Na+K (mEq)	Na+K / Na+K+Ca+Mg % (mEq)	Water type	Na+K / Na+K+Ca+Mg+Sr % (mEq)	CO_3 + HCO_3	SO_4	Cl
ANP 1	435356112420001	54	15	3.93	91.83	–	–	6.8	2.1	0.35	8.17	A	–	204	32	9.5
ANP 2	435100112420701	55	15	4.03	91.49	–	–	7.2	2.4	0.37	8.51	A	–	200	32	10
		55	17	4.19	86.54	–	–	13	3.4	0.65	13.46	A	–	210	41	14
		51	16	3.86	90.32	–	–	8.1	2.4	0.41	9.68	A	–	200	33	12
		55	15	4.03	91.4	–	–	7.3	2.4	0.38	8.6	A	–	199	32	10
ANP 3	435353112423201	23	2.8	1.38	74.44	–	–	9	3.2	0.47	25.56	D	–	18	42	8.2
ANP 5	435308112454101	43	17	3.54	92.08	–	–	6.3	1.2	0.3	7.92	A	–	184	27	7
ANP 6	435152112443101	45.1	16.7	3.62	88.29	0.216	3.63	9.7	2.3	0.48	11.71	A	11.7	179	33.5	16.3
		45.2	17.2	3.72	88.54	0.218	3.73	9.6	2.5	0.48	11.46	A	11.44	180	32.2	17.2
		44	17	3.59	87.87	–	–	10	2.4	0.5	12.13	A	–	176	35	14
ANP 7	435522112444201	45	16	3.61	92.53	–	–	6	1.2	0.29	7.47	A	–	184	28	6.8
ANP 8	434952112411301	45	15	3.53	90.31	–	–	7	2.9	0.38	9.69	A	–	184	29	7.2
ANP 9	434856112400001	33.1	15.7	3.19	82.66	0.208	3.2	13.7	2.9	0.67	17.34	B	17.32	177	29.1	12.6
		37	15	3.08	81.74	–	–	14	3.1	0.69	18.26	B	–	168	30	11
ANP 10	434909112400401	34	15	2.93	81.91	–	–	13	3.2	0.65	18.09	B	–	164	28	9.5
Arbor Test 1	433509112384801	33.7	10.9	2.58	77.75	0.116	2.58	15.2	3	0.74	22.25	B	22.23	159	12.5	14.3
		34.9	11.5	2.69	79.21	0.125	2.69	14.4	3.1	0.71	20.79	B	20.78	162	12.4	14.5
		30	9.1	2.25	76.88	–	–	14	2.6	0.68	23.12	B	–	144	10	8.8
Arco City Well 4	433758112181701	53.5	13.5	3.78	93.55	0.257	3.79	5.4	1	0.26	6.45	A	–	209	19.9	6.5
Area 2	433223112470201	34.2	13.8	2.84	80.04	0.152	2.85	14.3	3.4	0.71	19.96	B	19.94	170	16.8	17.3
		31	12	2.53	78.71	–	–	14	3	0.69	21.29	B	–	148	15	13
Atomic City	432638112484101	34.1	13.4	2.8	79.43	0.164	2.81	14.7	3.4	0.73	20.57	B	20.55	168	16.1	17.2
BFW	433042112535101	37.9	14.1	3.05	86.39	0.197	3.06	9.7	2.3	0.48	13.61	A	13.59	162	21.4	16.9
CFA 1	433204112562001	53.8	21	4.66	84.61	0.37	4.67	17.2	3.9	0.85	15.39	B	15.36		39.8	92.5
CFA 2	433144112563501	64.6	18.6	4.6	86.67	0.37	4.61	14.4	3.2	0.71	13.33	C/A	13.31	160	27.7	74
		71.9	26.4	5.76	84.7	0.483	5.77	21.4	4.3	1.04	15.3	C/B	15.28	149	45	115
		39	15	3.18	88.01	–	–	7.9	3.5	0.43	11.99	C/A	–	150	21	24
		42	15	3.33	88.68	–	–	8.3	2.5	0.42	11.32	A	–	162	21	16
		42	13	3.17	88	–	–	8.4	2.6	0.43	12	C/A	–	149	21	23
CPP 1	433433112560201	43	13	3.22	88.13	–	–	7.9	3.5	0.43	11.87	A	–	174	20	10
		43	12	3.38	88.84	–	–	8.3	2.5	0.42	11.16	A	–	187	22	9
CPP 2	433432112560801	44	13	3.27	89.82	–	–	7.8	1.2	0.37	10.18	A	–	176	21	10
CPP 3	433413112560401	39	12	2.93	88.8	–	–	7.8	1.2	0.37	11.2	A	–	160	21	9.8

Appendix D. Cation and anion data for selected wells at the Idaho National Laboratory, Idaho.—Continued

[**Local name:** is local well identifier used in this study. Concentrations are in milligrams per liter. **Site identifier:** is the unique numerical identifiers used to access well data (http://waterdata.usgs.gov/nwis). Wells with compositional and temporal variations in water type as noted by Olmsted (1962) are indicated in bold. **Water type:** is based on criteria used by Olmsted (1962). **Abbreviations:** Ca, calcium; Mg, magnesium; Sr, strontium; Na, sodium; K, potassium; CO₃, carbonate; HCO₃, bicarbonate; SO₄, sulfate; Cl, chloride; NO₃, nitrate; F, fluoride; mEq, milliequivalents; %, percent; –, not sampled]

Local name	Site identifier	Ca	Mg	Ca+Mg (mEq)	Ca+Mg % (mEq)	Sr	Ca+Mg+Sr (mEq)	Na	K	Na+K (mEq)	Na+K / Na+K+Ca+Mg % (mEq)	Water type	Na+K / Na+K+Ca+Mg+Sr % (mEq)	CO₃ + HCO₃	SO₄	Cl
EBR 1	433051113002601	22.6	15.3	2.39	84.82	0.195	2.39	8	3.1	0.43	15.18	**B**	15.16	144	15.7	7
		22	16	2.41	82.9	–	–	9.1	4	0.5	17.1	**B**	–	140	18	6
		22	15	2.33	82.74	–	–	9.3	3.2	0.49	17.26	**B**	–	140	18	6.2
		22	16	2.41	84.44	–	–	8.7	2.6	0.44	15.56	**B**	–	140	17	6.3
		22	16	2.41	82.3	–	–	9	5	0.52	17.7	**B**	–	144	18	6.2
		23	15	2.38	83.6	–	–	8.8	3.3	0.47	16.4	**B**	–	142	16	6
		23	16	2.46	84.79	–	–	8.4	3	0.44	15.21	**B**	–	140	17	6.2
EBR II-1	433546112391601	32	9.7	2.4	77.74	–	–	14	3	0.69	22.26	B	–	149	13	12
Engberson Well (ML-9)	435028112264501	68.7	28.6	5.78	76.16	0.331	5.79	38.2	5.8	1.81	23.84	C/B	23.82	281	42.1	55
EOCR 1	433120112535101	45	18	3.73	89.91	–	–	8.5	1.9	0.42	10.09	A	–	188	22	15
ETR	433521112574201	52	18	4.08	90.37	–	–	9.1	1.5	0.43	9.63	A	–	214	25	13
FET 1	435120112432101	51	15	3.78	89.8	–	–	7.4	4.2	0.43	10.2	A	–	182	33	11
FET 2	435119112431801	46	15	3.53	88.9	–	–	7.9	3.8	0.44	11.1	A	–	178	31	9
FET Disp 3	435124112433701	52	14	3.75	90.45	–	–	7.8	2.2	0.4	9.55	A	–	176	34	12
Fire Station 2	433548112562301	54.8	17.8	4.2	91.03	0.303	4.21	8.1	2.4	0.41	8.97	A	8.95	204	23.5	17.6
GCRE 1	433156112494401	51	15	3.78	91.25	–	–	7.1	2.1	0.36	8.75	A	–	188	22	14
IET 1	435153112420501	35	15	2.98	79.31	–	–	16	3.2	0.78	20.69	B	–	158	24	18
IET 1 Disposal	435153112420501	41	14	3.2	81.22	–	–	15	3.4	0.74	18.78	C/B	–	156	39	16
		49	13.9	3.59	82.08	0.251	3.59	15.9	3.6	0.78	17.92	B	17.9	189	29.9	18.7
INEL-1 WS	433717112563501	67.5	27.4	5.62	88.97	0.309	5.63	14.5	2.6	0.7	11.03	C/A	11.02	195	40.4	66.6
Leo Rogers 1	432533112504901	39.6	14.3	3.15	79.33	0.145	3.16	17	3.2	0.82	20.67	B	20.65	171	18.1	18.8
LPTF Disp	434946112412401	43	15	3.38	89.7	–	–	7.1	3.1	0.39	10.3	A	–	178	29	6.5
MTR Test	433520112572601	52	19	4.16	95.99	–	–	2.7	2.2	0.17	4.01	A	–	216	20	12
		52	19	4.16	96.14	–	–	2.6	2.1	0.17	3.86	A	–	216	20	11
MTR 1	433521112573801	48	19	3.96	88.72	–	–	8.4	5.4	0.5	11.28	A	–	212	21	13
		55	16	4.06	91.1	–	–	8.3	1.4	0.4	8.9	A	–	211	23	12
MTR 2	433523112575001	36	16	3.11	87.74	–	–	8	3.4	0.43	12.26	A	–	176	18	8.8
		36	16	3.11	87.34	–	–	7.9	4.2	0.45	12.66	A	–	176	18	7.5
		39	18	3.43	90.02	–	–	7.5	2.1	0.38	9.98	A	–	183	19	10
		40	17	3.39	88.69	–	–	8.6	2.3	0.43	11.31	A	–	183	22	10
Neville Well (ML-7)	435540112092901	32.4	10.5	2.48	76.37	0.092	2.48	16	2.8	0.77	23.63	B	23.61	148	14.7	11.7
NRF 1	433859112545401	69	21	5.17	90.83	–	–	9	5.1	0.52	9.17	C/A	–	224	35	35
		67	21	5.07	91.92	–	–	9.6	1.1	0.45	8.08	C/A	–	224	35	35
		66	22	5.1	91.26	–	–	10	2.1	0.49	8.74	C/A	–	223	36	36
NRF 2	433854112545401	68	21	5.12	91.59	–	–	9.4	2.4	0.47	8.41	C/A	–	226	36	38
		67	20	4.99	91.12	–	–	10	2	0.49	8.88	C/A	–	225	38	33
NRF 3	433858112545501	69	21	5.17	91.41	–	–	10	2	0.49	8.59	C/A	–	223	37	36

Appendix D. Cation and anion data for selected wells at the Idaho National Laboratory, Idaho.—Continued

[**Local name:** is local well identifier used in this study. Concentrations are in milligrams per liter. **Site identifier:** is the unique numerical identifiers used to access well data (http://waterdata.usgs.gov/nwis). Wells with compositional and temporal variations in water type as noted by Olmsted (1962) are indicated in bold. **Water type:** is based on criteria used by Olmsted (1962). **Abbreviations:** Ca, calcium; Mg, magnesium; Sr, strontium; Na, sodium; K, potassium; CO$_3$, carbonate; HCO$_3$, bicarbonate; SO$_4$, sulfate; Cl, chloride; NO$_3$, nitrate; F, fluoride; mEq, milliequivalents; %, percent; –, not sampled]

Local name	Site identifier	Ca	Mg	Ca+Mg (mEq)	Ca+Mg % (mEq)	Sr	Ca+Mg+Sr (mEq)	Na	K	Na+K (mEq)	Na+K / Na+K+Ca+Mg % (mEq)	Water type	Na+K / Na+K+Ca+Mg+Sr % (mEq)	CO$_3$ + HCO$_3$	SO$_4$	Cl
NPR Test	43344911252523101	51.4	14.3	3.74	90.84	0.242	3.75	7.5	2	0.38	9.16	A	9.15	196	23.1	14.7
		43.1	14.1	3.61	90.78	0.244	3.62	7.2	2.1	0.37	9.22	A	9.21	195	20.9	13.6
OMRE	43311611253470l	44	15	3.43	89.62	–	–	7.9	2.1	0.4	10.38	A	–	182	20	12
Pancheri 6	43572811310370l	43.6	14.5	3.37	92.45	0.135	3.37	5.8	0.9	0.28	7.55	A	7.55	186	15.5	10.1
Park Bell (08N 34E 17CCC7)	440058112293605	25.1	5	1.66	59.97	0.078	1.67	22.6	5	1.11	40.03	B	40.01	155	10.2	6.1
PSTF Test	43494111245420l	33.4	14.9	2.74	88.85	0.132	2.75	6.5	2.4	0.34	11.15	A	11.14	164	14.4	6.6
		33	15	2.88	88.19	–	–	7.4	2.5	0.39	11.81	A	–	160	16	7
P&W 1	43541611246040l	45	16	3.56	92.12	–	–	6.3	1.2	0.3	7.88	A	–	182	28	6.5
P&W 2	43541911245310l	33	14.3	3.07	89.59	0.137	3.08	7.5	1.2	0.36	10.41	A	10.4	171	25.8	5.5
		33.6	15.2	3.23	90.26	0.164	3.23	7.3	1.2	0.35	9.74	A	9.73	170	24.7	7.5
		43	16	3.46	90.78	–	–	7.2	1.5	0.35	9.22	A	–	178	30	7.5
P&W 3	43544311243580l	47	16	3.66	92.46	–	–	6.1	1.3	0.3	7.54	A	–	183	28	7.5
RWMC M3S	43500811302180l	43.4	15	3.4	88.93	0.244	3.41	8.2	2.6	0.42	11.07	A	11.05	176	24.3	13.4
RWMC M7S	43502311301480l	37.6	14	3.13	88.45	0.231	3.13	7.8	2.7	0.41	11.55	A	11.53	172	22.2	11.9
Site 4	43561711254200l	45.3	14.1	3.42	89.88	0.231	3.43	7.8	1.8	0.39	10.12	A	10.11	192	19.4	10.1
		47	13	3.42	91.13	–	–	6.7	1.6	0.33	8.87	A	–	181	21	8.5
Site 6-1	43382611251070l	44	14	3.2	90.1	–	–	6.9	2	0.35	9.9	A	–	168	19	9
Site 9	43312311253010l	35.7	14.7	2.99	84.26	0.188	3	11.2	2.8	0.56	15.74	B	15.72	166	22.9	13.1
Site 14	43434111246310l	34	13.3	2.79	81.4	0.215	2.8	12.9	3	0.64	18.6	**B**	18.58	165	23.4	9.5
Site 16	43354511239150l	31	10	2.37	76.6	–	–	15	2.8	0.72	23.4	B	–	152	12	11
		33	9.7	2.44	76.78	–	–	15	3.4	0.74	23.22	B	–	149	12	10
Site 17	43402711257570l	51	17.3	3.97	89.62	0.206	3.97	9.8	1.3	0.46	10.38	A	10.37	219	20.4	9.9
Site 19	43352211258210l	42.4	17.5	3.56	89.97	0.211	3.56	8	1.9	0.4	10.03	A	10.02	200	20.7	11.6
		40	18	3.48	90.01	–	–	7.7	2	0.39	9.99	A	–	176	22	12
		42	18	3.58	90.36	–	–	7.6	2	0.38	9.64	A	–	186	21	12
SL-1	43310711249220l	33	13	2.72	79.85	–	–	14	3	0.69	20.15	B	–	151	18	16
SPERT 1	43325211252030l	39	14	3.1	87.27	–	–	8.8	2.7	0.45	12.73	A	–	158	19	16
SPERT 2	43247111251520l	35	13	2.82	81.67	–	–	13	2.6	0.63	18.33	B	–	160	23	10
TAN Exploration	43503811245340l	34	15.2	2.95	84.89	0.181	2.95	10	3.5	0.52	15.11	C/B	15.09	139	24.2	19.9
USGS 1	43270011247080l	31.2	11.9	2.54	79.13	0.133	2.54	13.5	3.2	0.67	20.87	B	20.85	158	13	13
		28	11	2.3	74.62	–	–	18	2.7	0.78	25.38	B	–	150	14	10
		28	11	2.3	77.25	–	–	14	2.7	0.68	22.75	B	–	148	14	9
USGS 2	43332011243230l	35.4	12.1	2.76	78.84	0.131	2.77	15.1	3.3	0.74	21.16	B	21.14	166	14.1	17
		32	12	2.58	84.85	–	–	8.9	2.9	0.46	15.15	B	–	149	14	8.8
		29	11	2.35	77.56	–	–	14	2.8	0.68	22.44	B	–	147	12	9
USGS 3A	43773211233540l	29	10	2.27	75.56	–	–	15	3.2	0.73	24.44	B	–	146	9.1	11
		29	9.5	2.23	75.55	–	–	15	2.7	0.72	24.45	B	–	149	9.1	8

Appendix D. Cation and anion data for selected wells at the Idaho National Laboratory, Idaho.—Continued

[Local name: is local well identifier used in this study. Concentrations are in milligrams per liter. Site identifier: is the unique numerical identifiers used to access well data (http://waterdata.usgs.gov/nwis). Wells with compositional and temporal variations in water type as noted by Olmsted (1962) are indicated in bold. Water type: is based on criteria used by Olmsted (1962). Abbreviations: Ca, calcium; Mg, magnesium; Sr, strontium; Na, sodium; K, potassium; CO_3, carbonate; HCO_3, bicarbonate; SO_4, sulfate; Cl, chloride; NO_3, nitrate; F, fluoride; mEq, milliequivalents; %, percent; –, not sampled]

Local name	Site identifier	Ca	Mg	Ca+Mg (mEq)	Ca+Mg % (mEq)	Sr	Ca+Mg+Sr (mEq)	Na	K	Na+K (mEq)	Na+K / Na+K+Ca+Mg % (mEq)	Water type	Na+K / Na+K+Ca+Mg+Sr % (mEq)	CO_3 + HCO_3	SO_4	Cl
USGS 4	43465711282201	74.4	25.8	5.84	74.68	0.322	5.84	41.9	6.1	1.98	25.32	B	25.3	343	34.1	42.4
		68.9	26	5.58	75.37	0.312	5.58	37.9	6.8	1.82	24.63	B	24.6	340	31.9	40.4
		100	36	7.95	79.49	–	–	42	8.8	2.05	20.51	C/B	–	185	59	174
		93	33	7.36	78.62	–	–	42	6.8	2	21.38	C/B	–	186	57	160
USGS 5	43354311249380 1	39.8	12.6	3.02	89.36	0.189	3.03	7.1	2	0.36	10.64	A	10.63	171	18.7	9.4
		24	8	1.86	77.14	–	–	10	4.5	0.55	22.86	C/B	–	74	23	10
		41	13	3.12	89.73	–	–	7.2	1.7	0.36	10.27	A	–	172	20	9.5
USGS 6	43403111245370 1	28.7	11.3	2.36	80.55	0.182	2.37	11.7	2.4	0.57	19.45	B	19.42	146	18.1	9.4
		76	21	5.52	84.56	–	–	21	3.7	1.01	15.44	C/B	–	102	37	138
		30	12	2.48	78.56	–	–	14	2.7	0.68	21.44	B	–	150	20	9.5
USGS 7	43491511244390 1	24.6	9.3	1.99	66.2	0.12	2	20.8	4.4	1.02	33.8	B	33.76	142	16.1	9.1
		32	16	2.91	76.96	–	–	17	5.2	0.87	23.04	B	–	167	23	10
		34	15	2.93	79.08	–	–	15	4.8	0.78	20.92	B	–	173	20	10
		33	16	2.96	83.13	–	–	12	3.1	0.6	16.87	B	–	170	20	8
		26	11	2.2	65.77	–	–	24	4	1.15	34.23	B	–	148	21	9.5
		23	7.7	1.78	56.92	–	–	31	4	1.35	43.08	B	–	147	20	10
		11	8.2	1.22	48.59	–	–	27	4.7	1.29	51.41	B	–	118	1.4	10
		21	8.8	1.77	59.62	–	–	25	4.4	1.2	40.38	B	–	142	7.4	10
		22	8.4	1.79	59.85	–	–	25	4.4	1.2	40.15	B	–	144	8.2	10
		23	7.6	1.77	59.59	–	–	25	4.5	1.2	40.41	B	–	146	9.6	10
USGS 8	43312111311580 1	46.8	15	3.57	91.16	0.254	3.58	6.9	1.8	0.35	8.84	A	8.83	201	21	8.4
		44	15	3.43	91.59	–	–	6.3	1.6	0.31	8.41	A	–	180	24	10
USGS 9	43274011304450 1	39.7	15.1	3.22	84.32	0.189	3.23	11.9	3.2	0.6	15.68	B	15.66	168	26.9	19.4
		40.7	15.6	3.31	84.24	0.202	3.32	12.2	3.5	0.62	15.76	B	15.74	172	20.9	26
		37	14	3	86.41	–	–	7.9	5	0.47	13.59	A	–	162	22	10
USGS 11	43233611306420 1	41.2	14.2	3.22	88.71	0.22	3.23	8.2	2.1	0.41	11.29	A	11.27	173	23.1	11.8
		40.7	14.5	3.22	88.8	0.234	3.23	8	2.3	0.41	11.2	A	11.19	177	22	11.8
		38	13	2.97	89	–	–	6.9	2.6	0.37	11	A	–	162	21	8.8
		38	14	3.05	90.13	–	–	6.5	2	0.33	9.87	A	–	158	24	7.4
USGS 12	43412611255070 1	71.1	23.3	5.47	87.96	0.357	5.47	15.9	2.2	0.75	12.04	C/A	12.02	261	37	37.6
		70	22	5.3	91.77	–	–	10	1.6	0.48	8.23	C/A	–	221	40	44
		67	22	5.15	90.68	–	–	11	2	0.53	9.32	C/A	–	218	38	40
USGS 13	43273111314390 2	34	21	3.42	72.76	–	–	27	4.2	1.28	27.24	C/B	–	151	29	55
USGS 14	43201911256320 1	36.9	15.3	3.1	80.61	0.178	3.1	15.5	2.8	0.75	19.39	B	19 37	168	21.5	21
USGS 15	43423411255170 1	32.7	15.3	2.89	87.1	0.188	2.9	8.9	1.6	0.43	12.9	A	12.88	167	18.5	7
		34.8	16	3.05	87.27	0.191	3.06	9.3	1.6	0.45	12.73	A	12.72	171	19.9	10
		5.7	3.1	0.54	14.57	–	–	70	4.6	3.16	85.43	B	–	154	3.6	17
		32	16	2.91	87.94	–	–	8.3	1.5	0.4	12.06	A	–	163	19	7

Appendix D. Cation and anion data for selected wells at the Idaho National Laboratory, Idaho.—Continued

[Local name: is local well identifier used in this study. Concentrations are in milligrams per liter. Site identifier: is the unique numerical identifiers used to access well data (http://waterdata.usgs.gov/nwis). Wells with compositional and temporal variations in water type as noted by Olmsted (1962) are indicated in bold. Water type: is based on criteria used by Olmsted (1962). Abbreviations: Ca, calcium, Mg, magnesium; Sr, strontium; Na, sodium; K, potassium; CO$_3$, carbonate; HCO$_3$, bicarbonate; SO$_4$, sulfate; Cl, chloride; NO$_3$, nitrate; F, fluoride; mEq, milliequivalents; %, percent; –, not sampled]

Local name	Site identifier	Ca	Mg	Ca+Mg (mEq)	Ca+Mg % (mEq)	Sr	Ca+Mg+Sr (mEq)	Na	K	Na+K (mEq)	Na+K / Na+K+Ca+Mg % (mEq)	Water type	Na+K / Na+K+Ca+Mg+Sr % (mEq)	CO$_3$ + HCO$_3$	SO$_4$	Cl
USGS 17	433937112515401	37.4	10.1	2.7	88.26	0.208	2.7	6.9	2.3	0.36	11.74	A	11.73	151	19.1	4.9
		34	9.8	2.5	86.78	–	–	7	3	0.38	13.22	A	–	143	16	6.8
		27	6.6	1.89	82.25	–	–	7.5	3.2	0.41	17.75	**B**	–	100	21	7
		23	5.9	1.88	82.05	–	–	7	4.2	0.41	17.95	C/B	–	81	21	6.6
USGS 18	434540112440901	35.1	15.8	3.05	83.56	0.165	3.06	12.1	2.9	0.6	16.44	B	16.42	168	24.7	10.2
		33	16	2.96	81.44	–	–	13	4.3	0.68	18.56	B	–	166	27	9.2
USGS 19	434426112575701	43.8	16.4	3.54	87.11	0.249	3.54	11.2	1.4	0.52	12.89	A	12.87	205	22.3	8.9
		44.1	16.9	3.59	87.88	0.256	3.6	10.5	1.5	0.5	12.12	A	12 1	197	20.6	9.9
		47	17	3.74	91.16	–	–	7	2.3	0.36	8.84	A	–	186	26	14
USGS 20	433253112545901	27	10	2.17	50.45	–	–	46	5.1	2.13	49.55	C/B	–	140	25	53
		35	11	2.65	86.82	–	–	7.9	2.3	0.4	13.18	A	–	144	17	12
USGS 21	434307112382601	33	11	2.55	77.41	–	–	15	3.6	0.74	22.59	B	–	140	25	14
USGS 22	434342113031701	34.7	10.6	2.6	71.13	0.114	2.61	21	5.6	1.06	28.87	C/B	28.85	87	21	66.5
		31	10	2.37	71.02	–	–	20	3.8	0.97	28.98	C/B	–	88	22	47
USGS 23	434055112595901	37.4	15.8	3.17	87.77	0.219	3.17	9.2	1.6	0.44	12.23	A	12 21	182	17.6	9.9
		39	18	3.43	90.45	–	–	7.5	1.4	0.36	9.55	A	–	176	20	11
USGS 24	435053112420801	51	13	3.61	91.27	–	–	6.6	2.3	0.35	8.73	A	–	189	27	7.4
USGS 25	435339112444601	45	17	3.64	92.1	–	–	6.3	1.5	0.31	7.9	A	–	187	29	6.5
USGS 26	435212112394001	41.2	14.9	3.28	83.18	0.196	3.29	13.2	3.5	0.66	16.82	B	16.8	182	28.6	13.3
		39	15	3.18	81.08	–	–	15	3.5	0.74	18.92	B	–	179	30	11
USGS 27	434851112321801	54	19.1	4.27	77.94	0.256	4.27	24	6.4	1.21	22.06	C/B	22.04	170	38.5	61.9
USGS 28	434600112360101	33	14	2.8	78.99	–	–	15	3.6	0.74	21.01	B	–	164	25	9
USGS 29	434407112285101	48.5	14.1	3.58	79.67	0.158	3.58	19	3.4	0.91	20.33	B	20.31	192	16.7	26
		46.7	13.8	3.47	79.89	0.15	3.47	18	3.5	0.87	20.11	B	20.09	195	16	27.2
USGS 30		26	13	2.37	76.26	–	–	15	3.3	0.74	23.74	C/B	–	100	19	32
		26	13	2.37	74	–	–	17	3.6	0.83	26	**C/B**	–	128	28	14
		36	13	2.87	76.35	–	–	18	4.1	0.89	23.65	**B**	–	172	27	14
USGS 31	434625112342101	42.1	15	3.34	81.98	0.188	3.34	14.8	3.5	0.73	18.02	C/B	18	169	28.7	21.2
		41	15.3	3.31	81.99	0.196	3.31	14.4	3.9	0.73	18.01	C/B	17 99	172	27.7	22.5
		36	13	2.87	80.52	–	–	14	3.3	0.69	19.48	B	–	169	24	10
USGS 32	434444112322101	49.5	18.7	4.01	82.05	0.246	4.01	17.7	4.2	0.88	17.95	C/B	17 93	163	39.1	42
		35	14	2.9	73.86	–	–	21	4.4	1.03	26.14	B	–	164	27	18
USGS 33	434314112322901	51	19	4.11	72.4	–	–	33	5.1	1.57	27.6	C/B	–	164	32	69
USGS 34	434334112565501	–	–	–	–	–	–	–	–	–	25	B	–	–	–	–
USGS 36	433301112565201	60.5	15.4	4.31	84.37	0.334	4.31	16.7	2.8	0.8	15.63	C/B	15.61	199	27.7	33.9
USGS 40	434111112561101	56	15	4.03	77.24	–	–	25	3.9	1.19	22.76	C/B	–	144	24	81
		55	13	3.81	–	–	–	84	3.9	3.75	49.6	C/B	–	188	34	126
USGS 41	434091112561301	56	13	3.86	–	–	–	80	4	3.58	48.11	C/B	–	189	34	126

Appendix D. Cation and anion data for selected wells at the Idaho National Laboratory, Idaho.—Continued

[Local name: is local well identifier used in this study. Concentrations are in milligrams per liter. Site identifier: is the unique numerical identifiers used to access well data (http://waterdata.usgs.gov/nwis). Wells with compositional and temporal variations in water type as noted by Olmsted (1962) are indicated in bold. Water type: is based on criteria used by Olmsted (1962). Abbreviations: Ca, calcium; Mg, magnesium; Sr, strontium; Na, sodium; K, potassium; CO_3, carbonate; HCO_3, bicarbonate; SO_4, sulfate; Cl, chloride; NO_3, nitrate; F, fluoride; mEq, milliequivalents; %, percent; –, not sampled]

Local name	Site identifier	Ca	Mg	Ca+Mg (mEq)	Ca+Mg % (mEq)	Sr	Ca+Mg+Sr (mEq)	Na	K	Na+K (mEq)	Na+K / Na+K+Ca+Mg % (mEq)	Water type	Na+K / Na+K+Ca+Mg+Sr % (mEq)	CO_3 + HCO_3	SO_4	Cl
USGS 42	433404112561301	54	14	3.85	71.75	–	–	33	3.1	1.51	28.25	C/B	–	190	25	54
		52	12	3.58	73.53	–	–	28	2.8	1.29	26.47	C/B	–	186	25	40
		54	11	3.6	65.29	–	–	42	3.4	1.91	34.71	C/B	–	196	27	56
		52	12	3.58	68.53	–	–	36	3.1	1.65	31.47	B	–	187	27	48
USGS 43	433415112561501	27	8.6	2.06	71.57	–	–	17	3	0.82	28.43	B	–	129	21	8
		54	11	3.6	58.82	–	–	56	3.3	2.52	41.18	C/B	–	186	29	77
		54	11	3.6	57.17	–	–	60	3.4	2.7	42.83	C/B	–	194	28	84
USGS 44	433409112562101	33	9.7	2.44	76.53	–	–	15	3.8	0.75	23.47	B	–	161	20	6.5
		49	14	3.6	89.43	–	–	8.6	2	0.43	10.57	A	–	188	22	11
		43	13	3.22	87.47	–	–	9.3	2.2	0.46	12.53	A	–	174	21	10
USGS 45	433402112561801	52	12	3.58	84.19	–	–	14	2.5	0.67	15.81	B	–	–	–	–
USGS 46	433407112561501	55	13	3.81	74.1	–	–	29	2.8	1.33	25.9	C/B	–	190	26	46
		50	12	3.48	81.3	–	–	17	2.4	0.8	18.7	B	–	192	21	21
USGS 47	433407112560301	52	12	3.58	69.14	–	–	35	3	1.6	30.86	C/B	–	194	23	48
		51	12	3.53	71.96	–	–	30	2.8	1.38	28.04	C/B	–	194	25	40
USGS 48	433401112560301	60	12	3.98	47.7	–	–	98	4	4.37	52.3	C/B	–	193	39	154
		61	17	4.44	60.66	–	–	64	3.8	2.88	39.34	C/B	–	189	32	120
USGS 49	433403112555401	50	12	3.48	57.94	–	–	56	3.6	2.53	42.06	C/B	–	200	30	68
		54	12	3.68	62.76	–	–	48	3.8	2.19	37.24	C/B	–	190	29	73
		54	11	3.6	59.54	–	–	54	3.8	2.45	40.46	C/B	–	192	29	78
		54	11	3.6	60.86	–	–	51	3.8	2.32	39.14	C/A	–	192	27	76
USGS 51	433350112560601	55	14	3.9	86.68	–	–	12	3	0.6	13.32	C/A	–	169	23	40
USGS 55	433508112573001	160	31	10.54	74.21	PW	–	82	3.7	3.66	25.79	D/B	–	175	464	46
USGS 57	433344112562601	27.8	13.5	2.5	47.03	0.322	2.51	61.8	4.9	2.81	52.97	B	52.89	–	33.1	110.5
		23.5	11.8	2.14	49.62	0.265	2.15	47.8	3.8	2.18	50.38	B	50.31	–	28	73.5
USGS 65	433447112574501	55	14	3.9	91.3	–	–	7.3	2.1	0.37	8.7	C/A	–	149	60	16
USGS 66	433436112564801	35	21	3.47	75.11	–	–	24	4.2	1.15	24.89	C/B	–	194	44	16
USGS 77	433315112560301	70.3	18.8	5.06	75.44	0.425	5.06	34.9	5	1.65	24.56	B	24.53	–	36.6	133.4
USGS 82	433401112551001	35.7	13.4	2.88	84.5	0.214	2.89	10.4	3	0.53	15.5	B	15.48	152	21	18
USGS 83	433023112561501	27.3	10.6	2.23	82.14	0.157	2.24	9.7	2.5	0.49	17.86	B	17.84	123	20.1	10.8
USGS 86	432935113080001	37	10.2	2.69	82.93	0.171	2.69	11	2.9	0.55	17.07	C/B	17.04	132	22.7	19.6
USGS 89	433005113032801	27.2	15.6	2.64	75.15	0.118	2.64	17.9	3.7	0.87	24.85	C/B	24.83	103	34.9	38.8
USGS 97	433807112551501	73	24.3	5.64	88.6	0.316	5.65	15.4	2.2	0.73	11.4	C/A	11.39	269	35.9	38
USGS 98	433657112563601	48.9	18.3	3.95	88.88	0.222	3.95	10	2.3	0.49	11.12	A	11.11	209	21.7	15.2
USGS 99	433705112552101	59.8	22.6	4.84	89.4	0.252	4.85	12.2	1.7	0.57	10.6	A	10.59	247	27	22.2
USGS 100	433503112400701	38.3	12.3	2.92	79.04	0.133	2.93	16	3.1	0.78	20.96	B	20.94	164	21	17.7
USGS 101	433255112381801	36.9	12	2.83	79.88	0.166	2.83	14.5	3.2	0.71	20.12	B	20.1	169	14.8	16.4
		28.3	9.1	2.16	76.47	0.081	2.16	13.7	2.7	0.66	23.53	B	23.51	145	8.8	8.5
		28.8	9.2	2.19	77.62	0.111	2.2	12.9	2.8	0.63	22.38	B	22.36	148	9	8.5

Appendix D.   Cation and anion data for selected wells at the Idaho National Laboratory, Idaho.—Continued

[**Local name:** is local well identifier used in this study. Concentrations are in milligrams per liter. **Site identifier:** is the unique numerical identifiers used to access well data (http://waterdata.usgs.gov/nwis). Wells with compositional and temporal variations in water type as noted by Olmsted (1962) are indicated in bold. **Water type** is based on criteria used by Olmsted (1962). **Abbreviations:** Ca, calcium; Mg, magnesium; Sr, strontium; Na, sodium; K, potassium; CO_3, carbonate; HCO_3, bicarbonate; SO_4, sulfate; Cl, chloride; NO_3, nitrate; F, fluoride; mEq, milliequivalents; %, percent; –, not sampled]

Local name	Site identifier	Ca	Mg	Ca+Mg (mEq)	Ca+Mg % (mEq)	Sr	Ca+Mg+Sr (mEq)	Na	K	Na+K (mEq)	Na+K / Na+K+Ca+Mg % (mEq)	Water type	Na+K / Na+K+Ca+Mg+Sr % (mEq)	CO_3 + HCO_3	SO_4	Cl
USGS 102	433535112551601	75.9	23.2	5.6	89.69	0.308	5.6	13.5	2.2	0.64	10.31	A	10.3	264	35.5	34
USGS 103	432714112560701	36.2	14.8	3.02	82.66	0.183	3.03	13	2.7	0.63	17.34	B	17.32	167	24.2	17.1
	432714112560723	36.1	15.3	3.06	83.05	0.186	3.06	12.6	3	0.62	16.95	B	16.93	167	23.1	16.3
	432714112560720	2.6	14.6	2.28	76.01	0.148	2.28	14.7	3.1	0.72	23.96	C/B	23.96	123	23.1	18.4
	432714112560716	32.9	14.5	2.83	81.72	0.186	2.84	12.9	2.9	0.63	18.28	B	18.26	162	22.9	14.3
	432714112560712	30.7	13.7	2.66	84.75	0.172	2.66	9.3	2.8	0.48	15.25	B	15.23	158	19.6	10.2
	432714112560708	35.6	15.3	3.04	87.21	0.193	3.04	8.7	2.6	0.45	12.79	A	12.78	172	21.4	12.6
	432714112560704	37.1	14.9	3.08	87.65	0.191	3.08	8.5	2.5	0.43	12.35	A	12.33	176	22.3	14.3
	432714112560702	38.9	15.5	3.22	87.66	0.2	3.22	8.9	2.6	0.45	12.34	A	12.33	176	22.4	14.1
USGS 104	432356112560801	38.3	15.4	3.18	87.65	0.203	3.18	8.8	2.5	0.45	12.35	A	12.33	176	22.5	14.3
		35.4	13.5	2.88	87.43	0.19	2.88	8.1	2.4	0.41	12.57	A	12.55	154	19.7	11.8
		34.9	13.7	2.87	87.98	0.184	2.87	7.6	2.4	0.39	12.02	A	12	156	19.3	12.6
USGS 105	432703113001801	40.8	15.2	3.29	84.04	0.235	3.29	12.7	2.8	0.62	15.96	B	15.93	180	26	13.6
	432703113001818	33.6	14.3	3.1	84.22	0.207	3.11	11.7	2.8	0.58	15.78	B	15.76	–	–	–
	432703113001815	33.3	14.1	3.12	84.67	0.218	3.13	11.4	2.7	0.57	15.33	B	15.31	–	–	–
	432703113001811	33.8	14.8	3.2	85.13	0.216	3.21	11.2	2.8	0.56	14.87	A	14.85	–	–	–
	432703113001807	40	14.5	3.19	84.92	0.217	3.19	11.3	2.9	0.57	15.08	B	15.06	–	–	–
	432703113001803	39.9	14.2	3.16	85.3	0.217	3.16	10.9	2.8	0.54	14.7	A	14.69	176	25.3	21.3
USGS 107	432942112532801	37.6	16.6	3.24	81.02	0.12	3.25	15.4	3.5	0.76	18.98	B	18.96	165	22.4	14
USGS 108	432659112582601	37	15	3.08	85.5	0.189	3.09	10.6	2.4	0.52	14.5	A	14.48	176	26.9	19.4
USGS 109	432701113025601	41.2	16.2	3.39	85.71	0.228	3.39	11.4	2.7	0.56	14.29	A	14.27	181	25	14
		33.8	15.7	3.28	86.18	0.307	3.29	10.5	2.7	0.53	13.82	A	13.8	173	18	19
USGS 110A	432717112501502	35.7	14.9	3.06	80.23	0.205	3.06	15.2	3.6	0.75	19.77	B	19.74	173	29	151
USGS 112	433314112563001	75	21	5.52	69.05	0.484	5.53	54	4.9	2.47	30.95	C/B	30.9	164	31.2	218
USGS 113	433314112561801	73.3	23.1	5.81	61.94	0.532	5.82	78.4	6.2	3.57	38.06	C/B	38.01	146	21.2	38
USGS 115	433320112554101	42.9	13.3	3.24	83.09	0.244	3.24	13.2	3.3	0.66	16.91	C/B	16.89	122	34.2	89.3
USGS 116	433331112553201	55.4	16	4.13	77.54	0.333	4.14	24.8	4.6	1.2	22.46	C/B	22.42	121	17.1	13.7
USGS 117	432955113025901	25.7	11.2	2.2	81.99	0.144	2.21	9.6	2.6	0.48	18.01	B	17.99	186	38	21.7
USGS 120	432919113031501	34	18.4	3.21	72.68	0.197	3.22	25.4	4	1.21	27.32	C/B	27.3	172	22.4	14.3
USGS 124	433207112583101	39.3	16	3.28	88.08	0.191	3.28	8.9	2.2	0.44	11.92	A	11.9	176	21.5	14.8
USGS 125	433602113052801	38.6	16.2	3.26	88.01	0.256	3.27	8.8	2.4	0.44	11.99	A	11.97	178	25.8	14.9
		40.8	15.9	3.34	85.17	0.23	3.35	11.8	2.7	0.58	14.83	A	14.81	181	24.5	14.8
USGS 126A	435529112471301	39.8	15.8	3.29	85.7	0.285	3.29	10.9	2.9	0.55	14.3	A	14.27	–	–	–
USGS 126B	435529112471401	35.5	13.7	2.9	87.67	0.178	2.9	8.4	1.6	0.41	12.33	A	12.32	–	–	–
USGS 127	433058112572201	37.8	14.9	3.11	87.66	0.201	3.12	8.8	2.3	0.44	12.34	A	12.32	–	–	–
		32.8	12.3	2.65	85.99	0.198	2.65	8.3	2.8	0.43	14.01	A	13.99	–	–	–
USGS 128	433250112565601	53.8	13.9	3.83	74.45	0.315	3.84	28.3	3.2	1.31	25.55	C/B	25.51	–	–	–

Appendix D. Cation and anion data for selected wells at the Idaho National Laboratory, Idaho.—Continued

[**Local name:** is local well identifier used in this study. Concentrations are in milligrams per liter. **Site identifier:** is the unique numerical identifiers used to access well data (http://waterdata.usgs.gov/nwis). Wells with compositional and temporal variations in water type as noted by Olmsted (1962) are indicated in bold. **Water type:** is based on criteria used by Olmsted (1962). **Abbreviations:** Ca, calcium; Mg, magnesium; Sr, strontium; Na, sodium; K, potassium; CO₃, carbonate; HCO₃, bicarbonate; SO₄, sulfate; Cl, chloride; NO₃, nitrate; F, fluoride; mEq, milliequivalents; %, percent; –, not sampled]

Local name	Site identifier	Ca	Mg	Ca+Mg (mEq)	Ca+Mg % (mEq)	Sr	Ca+Mg+Sr (mEq)	Na	K	Na+K (mEq)	Na+K / Na+K+Ca+Mg % (mEq)	Water type	Na+K / Na+K+Ca+Mg+Sr % (mEq)	$CO_3 + HCO_3$	SO_4	Cl
USGS 132	432906113025022	35.9	18.1	3.28	72.71	0.24	3.29	26.1	3.8	1.23	27.29	C/B	27.26	177	40.8	35
	432906113025018	38.4	15.4	3.18	85.56	0.223	3.19	10.7	2.8	0.54	14.44	A	14.42	180	25.9	11.2
	432906113025014	39.6	14.5	3.17	87.53	0.222	3.17	8.9	2.6	0.45	12.47	A	12.45	178	24.8	10.3
	432906113025010	39.3	14.4	3.15	87.48	0.214	3.15	8.9	2.6	0.45	12.52	A	12.51	178	24.7	10.2
	432906113025006	39.9	14.6	3.19	87.28	0.218	3.2	9.2	2.6	0.47	12.72	A	12.7	178	24.8	10.3
	432906113025001	38.8	14.7	3.15	84.45	0.23	3.15	11.7	2.8	0.58	15.55	B	15 53	185	25.4	10.7
USGS 133	433605112554312	43.3	13.8	3.3	88.05	0.22	3.3	8.5	3	0.45	11.95	A	11 93	168	20.7	12.5
	433605112554308	39.8	13.2	3.07	88.48	0.266	3.08	7.6	2.7	0.4	11.52	A	11 5	174	21	12.9
	433605112554305	40.2	13.6	3.13	89.64	0.184	3.13	7.1	2.1	0.36	10.36	A	10.34	177	20.5	12.5
	433605112554301	50.4	17.8	3.98	90.15	0.217	3.98	9	1.8	0.43	9.85	A	9.84	229	22.1	13.9
USGS 134	433611112595819	24.8	16	2.55	87.36	0.145	2.56	7.2	2.2	0.37	12.64	A	12.62	156	–	–
	433611112595815	36.3	20.1	3.47	88.79	0.174	3.47	8.7	2.3	0.44	11.21	A	11 2	212	–	–
	433611112595811	25.5	15.5	2.55	86.66	0.153	2.55	7.7	2.2	0.39	13.34	A	13.32	158	–	–
	433611112595807	31.9	16.6	2.96	87.65	0.182	2.96	8.2	2.4	0.42	12.35	A	12.33	166	–	–
	433611112595804	25.2	15.6	2.54	86.27	0.131	2.54	7.8	2.6	0.4	13.73	A	13.72	158	19.1	7.5
USGS 135	432753113093613	42.4	11.7	3.08	90.19	0.237	3.08	6.7	1.8	0.33	9.81	A	9.79	–	–	–
	432753113093609	40.6	11.9	3.01	90.06	0.228	3.01	6.5	1.9	0.33	9.94	A	9 92	–	–	–
	432753113093605	42.9	12.4	3.16	90.4	0.241	3.17	6.6	1.9	0.34	9.6	A	9 59	–	–	–
	432753113093601	39.1	13.3	3.05	90.29	0.246	3.05	6.4	1.9	0.33	9.71	A	9.7	–	–	–
MIDDLE 2050A	433409112570515	53.2	16.5	4.01	89.54	0.272	4.02	9.4	2.4	0.47	10.46	A	10.44	–	–	–
	433409112570512	42.8	14.2	3.3	89.9	0.23	3.31	7.2	2.3	0.37	10.1	A	10.09	–	–	–
	433409112570509	45.2	17.2	3.67	89.67	0.211	3.68	8.6	1.9	0.42	10.33	A	10 32	–	–	–
	433409112570506	41.4	16.4	3.42	89.91	0.202	3.42	7.8	1.7	0.38	10.09	A	10.08	–	–	–
	433409112570503	40	17.4	3.43	85.06	0.273	3.43	12.4	2.5	0.6	14.94	A	14.92	–	–	–
MIDDLE 2051	433217112004912	46.3	11.1	3.22	91.29	0.255	3.23	5.7	2.4	0.31	8.71	A	8.7	–	–	–
	433217112004909	43.7	15.6	3.46	89.22	0.235	3.47	8.3	2.3	0.42	10.78	A	10.76	–	–	–
	433217112004906	43.1	15.3	3.41	89.06	0.229	3.42	8.4	2.2	0.42	10.94	A	10.92	–	–	–
	433217112004903	35.7	17.1	3.19	89.17	0.206	3.19	7.5	2.4	0.39	10.83	A	10.82	–	–	–
	433217112004901	36.8	17.8	3.3	89.12	0.214	3.31	7.8	2.5	0.4	10.88	A	10.86	–	–	–

Appendix D. Cation and anion data for selected wells at the Idaho National Laboratory, Idaho.—Continued

[Local name: is local well identifier used in this study. Concentrations are in milligrams per liter. Site identifier: is the unique numerical identifiers used to access well data (http://waterdata.usgs.gov/nwis). Wells with compositional and temporal variations in water type as noted by Olmsted (1962) are indicated in bold. Water type: is based on criteria used by Olmsted (1962). Abbreviations: Ca, calcium; Mg, magnesium; Sr, strontium; Na, sodium; K, potassium; CO$_3$, carbonate; HCO$_3$, bicarbonate; SO$_4$, sulfate; Cl, chloride; NO$_3$, nitrate; F, fluoride; mEq, milliequivalents; %, percent; –, not sampled]

Local name	Site identifier	NO$_3$	F	Cl+NO$_3$+F	CO$_3$+HCO$_3$ (mEq)	SO$_4$ (mEq)	Cl (mEq)	NO$_3$ (mEq)	F (mEq)	CO$_3$+HCO$_3$ % (mEq)	SO$_4$ % (mEq)	Cl+NO$_3$+F % (mEq)	Cation anion balance	Reference
ANP 1	43505611242420001	2.5	0	11.8	3.34	0.67	0.27	0.04	0	77.49	15.44	7.07	-0.04	Olmsted (1962)
ANP 2	43510011242420701	3	0.2	13.2	3.28	0.67	0.28	0.05	0.01	76.49	15.55	7.96	0.12	Olmsted (1962)
		5	0.3	19.3	3.44	0.85	0.39	0.08	0.02	71.9	17.83	10.26	-0.06	Olmsted (1962)
		3.4	0.2	15.4	3.28	0.69	0.34	0.05	0.01	75.08	15.74	9.18	-0.09	Olmsted (1962)
		2.0	0.2	13.1	3.26	0.67	0.28	0.05	0.01	76.43	15.61	7.95	0.14	Olmsted (1962)
ANP 3	43505311242423201	1.2	0.5	9.9	0.6	0.87	0.23	0.02	0.03	34.78	–	–	–	Olmsted (1962)
ANP 5	43530811242454101	1.5	0.2	9.1	3.02	0.56	0.2	0.03	0.01	79.02	14.73	6.25	0.03	Olmsted (1962)
ANP 6	43515211242443101	3.6	0.24	20.1	2.93	0.7	0.46	0.06	0.01	70.49	16.76	12.75	-0.06	Busenberg and others (2000, table 7)
		3.7	0.26	21.2	2.95	0.67	0.49	0.06	0.01	70.59	16.04	13.36	0.02	Busenberg and others (2000, table 7)
ANP 7	43555211242444201	2.4	0.3	16.8	2.88	0.73	0.39	0.04	0.02	70.97	17.93	11.1	0.03	Olmsted (1962)
ANP 8	43495211242411301	1.6	0.2	8.6	3.02	0.58	0.19	0.03	0.01	78.8	15.23	5.96	0.08	Olmsted (1962)
		2.6	0.2	10	3.02	0.6	0.2	0.04	0.01	77.82	15.58	6.59	0.03	Olmsted (1962)
ANP 9	43485611242400001	3.7	0.39	16.1	2.9	0.61	0.36	0.05	0.02	73.76	15.41	10.83	-0.07	Busenberg and others (2000, table 7)
ANP 10	43490911242400401	1.7	0.4	13.1	2.75	0.62	0.31	0.03	0.02	73.68	16.72	9.6	0.03	Olmsted (1962)
		1.5	0.4	11.2	2.69	0.58	0.27	0.02	0.02	75.06	16.28	8.66	0	Olmsted (1962)
Arbor Test 1	43350911242384801	5.7	0.62	20.6	2.61	0.26	0.4	0.09	0.03	76.78	7.67	15.55	-0.08	Busenberg and others (2000, table 7)
		5.5	0.64	20.6	2.66	0.26	0.41	0.09	0.03	77.08	7.5	15.43	-0.05	Busenberg and others (2000, table 7)
Arco City Well 4	43375811213181701	2.4	0.8	12	2.36	0.21	0.25	0.04	0.04	81.46	7.19	11.36	0.02	Olmsted (1962)
		2.8	0.2	9.5	3.43	0.41	0.18	0.05	0.01	83.98	10.16	5.86	-0.04	Busenberg and others (2000, table 7)
Area 2	43322311242470201	4.8	0.44	22.5	2.79	0.35	0.49	0.08	0.02	74.81	9.39	15.8	-0.17	Busenberg and others (2000, table 7)
Atomic City	43263811242484101	2.5	0.5	15.8	2.43	0.31	0.37	0.04	0.03	76.57	9.86	13.58	0.05	Olmsted (1962)
		4.8	0.49	22.5	2.75	0.34	0.49	0.08	0.03	74.88	9.12	16	-0.15	Busenberg and others (2000, table 7)
BFW	43304211212535101	3	0.22	20.1	2.66	0.45	0.48	0.05	0.01	73	12.25	14.75	-0.1	Busenberg and others (2000, table 7)
CFA 1	43320411242562001	12.3	0.17	105.5	–	–	–	–	–	–	–	–	–	Busenberg and others (2000, table 7)
		15.4	0.25	89.7	2.62	0.58	2.09	0.25	0.01	47.27	10.4	42.34	-0.24	Busenberg and others (2000, table 7)

Appendix D. Cation and anion data for selected wells at the Idaho National Laboratory, Idaho.—Continued

[Local name: is local well identifier used in this study. Concentrations are in milligrams per liter. Site identifier: is the unique numerical identifiers used to access well data (http://waterdata.usgs.gov/nwis). Wells with compositional and temporal variations in water type as noted by Olmsted (1962) are indicated in bold. Water type: is based on criteria used by Olmsted (1962). Abbreviations: Ca, calcium; Mg, magnesium; Sr, strontium; Na, sodium; K, potassium; CO_3, carbonate; HCO_3, bicarbonate; SO_4, sulfate; Cl, chloride; NO_3, nitrate; F, fluoride; mEq, milliequivalents; %, percent; –, not sampled]

Local name	Site identifier	NO_3	F	$Cl+NO_3+F$	CO_3+HCO_3 (mEq)	SO_4 (mEq)	Cl (mEq)	NO_3 (mEq)	F (mEq)	CO_3+HCO_3 % (mEq)	SO_4 % (mEq)	$Cl+NO_3+F$ % (mEq)	Cation anion balance	Reference
CFA 2	433144112563501	**16.6**	**0.39**	**132**	**2.44**	**0.94**	**3.24**	**0.27**	**0.02**	**35.34**	**13.56**	**51.11**	**-0.11**	Busenberg and others (2000, table 7)
		1.9	0.2	26.1	2.46	0.44	0.68	0.03	0.01	68.03	12.1	19.87	0	Olmsted (1962)
		1.6	0.1	17.7	2.66	0.44	0.45	0.03	0.01	74.27	12.23	13.49	0.18	Olmsted (1962)
		1.4	0.2	24.6	2.44	0.44	0.65	0.02	0.01	68.57	12.28	19.15	0.04	Olmsted (1962)
CPP 1	433433112560201	0.9	0.2	11.1	2.85	0.42	0.28	0.01	0.01	79.76	11.65	8.59	0.07	Olmsted (1962)
		1.4	0.1	10.5	3.06	0.46	0.25	0.02	0.01	80.56	12.04	7.4	0	Olmsted (1962)
CPP 2	433432112560801	1	0.2	11.2	2.88	0.44	0.28	0.02	0.01	79.45	12.04	8.5	0	Olmsted (1962)
CPP 3	433413112560401	1.2	0.2	11.2	2.62	0.44	0.28	0.02	0.01	77.91	12.99	9.1	-0.06	Olmsted (1962)
EBR 1	433051113002601	1.6	0.19	8.8	2.36	0.33	0.2	0.03	0.01	80.82	11.19	7.99	-0.11	Busenberg and others (2000, table 7)
		1	0.2	7.2	2.29	0.37	0.17	0.02	0.01	80.08	13.08	6.84	0.05	Olmsted (1962)
		0.9	0.2	7.3	2.29	0.37	0.17	0.01	0.01	79.97	13.06	6.97	-0.05	Olmsted (1962)
		0.8	0.3	7.4	2.29	0.35	0.18	0.01	0.02	80.37	12.4	7.23	0	Olmsted (1962)
		1.3	0.4	7.9	2.36	0.37	0.17	0.02	0.02	79.95	12.7	7.35	-0.02	Olmsted (1962)
		0.8	0.1	6.9	2.33	0.33	0.17	0.01	0.01	81.72	11.7	6.58	0	Olmsted (1962)
		1.1	0.3	7.6	2.29	0.35	0.17	0.02	0.02	80.31	12.39	7.3	0.05	Olmsted (1962)
EBR II-1	433546112391601	1.9	0.7	14.6	2.44	0.27	0.34	0.03	0.04	78.3	8.68	13.02	-0.04	Olmsted (1962)
Engberson Well (ML-9)	435028112264501	23.5	0.19	78.7	4.61	0.88	1.55	0.38	0.01	62.05	11.81	26.14	0.17	Busenberg and others (2000, table 7)
EOCR 1	433120112535101	3	0.4	18.4	3.08	0.46	0.42	0.05	0.02	76.42	11.36	12.22	0.11	Olmsted (1962)
ETR	433521112574201	4.3	0.2	17.5	3.51	0.52	0.37	0.07	0.01	78.39	11.63	9.98	0.04	Olmsted (1962)
FET 1	435120112432101	3.1	0.1	14.2	2.98	0.69	0.31	0.05	0.01	73.92	17.03	9.06	0.17	Olmsted (1962)
FET 2	435119112431801	2.4	0.1	11.5	2.92	0.65	0.25	0.04	0.01	75.57	16.72	7.71	0.11	Olmsted (1962)
FET Disp 3	435124112433701	3.7	0.2	15.9	2.88	0.71	0.34	0.06	0.01	72.09	17.69	10.21	0.14	Olmsted (1962)
Fire Station 2	433548112562301	5.2	0.19	23	3.34	0.49	0.5	0.08	0.01	75.59	11.06	13.35	0.19	Busenberg and others (2000, table 7)
GCRE 1	433156112494401	2.7	0.1	16.8	3.08	0.46	0.39	0.04	0.01	77.36	11.5	11.14	0.16	Olmsted (1962)
		2.9	0.5	21.4	2.59	0.5	0.51	0.05	0.03	70.56	13.62	15.83	0.09	Olmsted (1962)
IET 1	435153112420501	2.6	0.3	18.9	2.56	0.81	0.45	0.04	0.02	65.93	20.94	13.13	0.06	Olmsted (1962)
IET 1 Disposal	435153112420501	6	0.21	24.9	3.1	0.62	0.53	0.1	0.01	71.12	14.29	14.59	0.02	Busenberg and others (2000, table 7)
INEL-1 WS	433717112563501	15.9	0.12	82.6	3.2	0.84	1.88	0.26	0.01	51.73	13.61	34.66	0.14	Busenberg and others (2000, table 7)
Leo Rogers 1	432533112504901	5	0.44	24.2	2.8	0.38	0.53	0.08	0.02	73.49	9.88	16.63	0.16	Busenberg and others (2000, table 7)
LPTF Disp	434946112412401	2.9	0.3	9.7	2.92	0.6	0.18	0.05	0.02	77.44	16.03	6.53	0	Olmsted (1962)
MTR Test	433520112572601	4.1	0.2	16.3	3.54	0.42	0.34	0.07	0.01	80.98	9.53	9.5	-0.04	Olmsted (1962)
		4.1	0.2	15.3	3.54	0.42	0.31	0.07	0.01	81.5	9.59	8.91	-0.02	Olmsted (1962)

Appendix D. Cation and anion data for selected wells at the Idaho National Laboratory, Idaho.—Continued

[**Local name:** is local well identifier used in this study. Concentrations are in milligrams per liter. **Site identifier:** is the unique numerical identifiers used to access well data (http://waterdata.usgs.gov/nwis). Wells with compositional and temporal variations in water type as noted by Olmsted (1962) are indicated in bold. **Water type:** is based on criteria used by Olmsted (1962). **Abbreviations:** Ca, calcium; Mg, magnesium; Sr, strontium; Na, sodium; k, potassium; CO_3, carbonate; HCO_3, bicarbonate; SO_4, sulfate; Cl, chloride; NO_3, nitrate; F, fluoride; mEq, milliequivalents; %, percent; –, not sampled]

Local name	Site identifier	NO_3	F	Cl+NO_3+F	CO_3+HCO_3 (mEq)	SO_4 (mEq)	Cl (mEq)	NO_3 (mEq)	F (mEq)	CO_3+HCO_3 % (mEq)	SO_4 % (mEq)	Cl+NO_3+F % (mEq)	Cation anion balance	Reference
MTR 1	43352111257380	0	0.2	13.2	3.47	0.44	0.37	0	0.01	81.01	10.19	8.79	0.17	Olmsted (1962)
		4.3	0.2	16.5	3.46	0.48	0.34	0.07	0.01	79.4	11	9.61	0.1	Olmsted (1962)
MTR 2	43352311257500	1.7	0	10.5	2.88	0.37	0.25	0.03	0	81.6	10.6	7.8	0.01	Olmsted (1962)
		1.7	0.2	9.4	2.88	0.37	0.21	0.03	0.01	82.21	10.68	7.11	0.06	Olmsted (1962)
		1.7	0.1	11.8	3	0.4	0.28	0.03	0.01	80.85	10.66	8.48	0.1	Olmsted (1962)
		2.3	0.1	12.4	3	0.46	0.28	0.04	0.01	79.31	12.11	8.58	0.05	Olmsted (1962)
Neville Well (ML-7)	43554011209290	15	0.81	27.5	2.43	0.31	0.33	0.24	0.04	72.49	9.15	18.37	-0.1	Busenberg and others (2000, table 7)
NRF 1	43385911254540	4.4	0.1	39.5	3.67	0.73	0.99	0.07	0.01	67.2	13.34	19.47	0.23	Olmsted (1962)
		3.8	0.2	39	3.67	0.73	0.99	0.06	0.01	67.25	13.35	19.4	0.06	Olmsted (1962)
		3.7	0.1	39.8	3.65	0.75	1.02	0.06	0.01	66.64	13.67	19.7	0.11	Olmsted (1962)
NRF 2	43385411254540	4.8	0.2	43	3.7	0.75	1.07	0.08	0.01	65.99	13.35	20.66	-0.02	Olmsted (1962)
		4.2	0.2	37.4	3.69	0.79	0.93	0.07	0.01	67.2	14.42	18.39	-0.01	Olmsted (1962)
NRF 3	43385811254550	1.5	0.2	38.1	3.65	0.77	1.02	0.03	0.01	66.67	14.05	19.27	0.18	Olmsted (1962)
NPR Test	43342912523101	3.4	0.19	18.3	3.21	0.48	0.41	0.05	0.01	76.98	11.53	11.49	-0.05	Busenberg and others (2000, table 7)
		3	0.22	16.8	3.2	0.44	0.38	0.05	0.01	78.43	10.68	10.89	-0.1	Busenberg and others (2000, table 7)
OMRE	43311611253470	2.1	0.2	14.3	2.98	0.42	0.34	0.03	0.01	73.87	11.01	10.12	-0.05	Olmsted (1962)
Pancheri 6	43572811310370	1.5	0.15	12.2	3.05	0.32	0.28	0.03	0.01	82.51	8.74	8.75	-0.05	Busenberg and others (2000, table 7)
Park Bell (08N 34E 17CCC7)	44005811229360		0.7	6.8	2.54	0.17	0.17	0	0.04	87.1	5.73	7.16	-0.14	Busenberg and others (2000, table 7)
PSTF Test	43492111245420	2.6	0.21	9.4	2.69	0.24	0.19	0.04	0.01	84.98	7.46	7.56	-0.08	Busenberg and others (2000, table 7)
P&W 1	43546112460401	2.5	0.1	9.4	2.62	0.33	0.2	0.04	0.01	82.07	10.43	7.51	0.07	Olmsted (1962)
P&W 2	43549112453101	2	0.3	8.8	2.98	0.58	0.18	0.03	0.02	73.55	15.35	6.09	0.07	Olmsted (1962)
		1.4	0.19	7.1	2.8	0.54	0.16	0.02	0.01	73.45	15.23	5.32	-0.1	Busenberg and others (2000, table 7)
		1.7	0.21	9.4	2.79	0.51	0.21	0.03	0.01	73.47	14.48	7.04	0.02	Busenberg and others (2000, table 7)
P&W 3	43544311243580	2.5	0.3	10.1	2.92	0.62	0.21	0.04	0.02	75.64	16.41	6.95	0.01	Olmsted (1962)
		1.7	0.1	9.3	3	0.58	0.21	0.03	0.01	75.38	15.24	6.38	0.13	Olmsted (1962)
RWMC M3S	43300811302180	3.5	0.3	17	2.88	0.51	0.38	0.05	0.02	75.17	13.18	11.65	-0.01	Busenberg and others (2000, table 7)
RWMC M7S	43303113014801	2.9	0.2	15	2.82	0.46	0.34	0.05	0.01	75.72	12.58	10.7	-0.14	Busenberg and others (2000, table 7)

Appendix D. Cation and anion data for selected wells at the Idaho National Laboratory, Idaho.—Continued

[Local name: is local well identifier used in this study. Concentrations are in milligrams per liter. Site identifier: is the unique numerical identifiers used to access well data (http://waterdata.usgs.gov/nwis). Wells with compositional and temporal variations in water type as noted by Olmsted (1962) are indicated in bold. Water type: is based on criteria used by Olmsted (1962). Abbreviations: Ca, calcium; Mg, magnesium; Sr, strontium; Na, sodium; K, potassium; CO_3, carbonate; HCO_3, bicarbonate; SO_4, sulfate; Cl, chloride; NO_3, nitrate; F, fluoride; mEq, milliequivalents; %, percent; –, not sampled]

Local name	Site identifier	NO_3	F	$Cl+NO_3+F$	CO_3+HCO_3 (mEq)	SO_4 (mEq)	Cl (mEq)	NO_3 (mEq)	F (mEq)	CO_3+HCO_3 % (mEq)	SO_4 % (mEq)	$Cl+NO_3+F$ % (mEq)	Cation anion balance	Reference
Site 4	433617112542001	2.5	0.2	12.8	3.15	0.4	0.28	0.04	0.01	80.97	10.39	8.64	-0.08	Busenberg and others (2000, table 7)
Site 6-1	433826112510701	1.6	0.2	10.3	2.97	0.44	0.24	0.03	0.01	80.61	11.88	7.5	0.07	Olmsted (1962)
Site 9	433123112530101	1.5	0.2	10.7	2.75	0.4	0.25	0.02	0.01	80.1	11.51	8.39	0.11	Olmsted (1962)
		2.7	0.28	16.1	2.72	0.48	0.37	0.04	0.01	75.05	13.15	11.8	-0.08	Busenberg and others (2000, table 7)
Site 14	434334112463101	2.5	0.41	12.4	2.7	0.49	0.27	0.04	0.02	76.8	13.84	9.37	-0.09	Busenberg and others (2000, table 7)
Site 16	433545112391501	2.1	0.7	13.8	2.49	0.25	0.31	0.03	0.04	79.79	8	12.2	-0.03	Olmsted (1962)
		2.2	0.7	12.9	2.44	0.25	0.28	0.04	0.04	80.16	8.2	11.63	0.14	Olmsted (1962)
Site 17	434027112575701	4.4	0.12	14.4	3.59	0.42	0.28	0.07	0.01	82.12	9.72	8.16	0.06	Busenberg and others (2000, table 7)
Site 19	433522112582101	4	0.19	15.8	3.28	0.43	0.33	0.06	0.01	79.74	10.49	9.77	-0.16	Busenberg and others (2000, table 7)
		3.4	0.3	15.7	2.88	0.46	0.34	0.05	0.02	76.89	12.21	10.9	0.11	Olmsted (1962)
		3.6	0.2	15.8	3.05	0.44	0.34	0.06	0.01	78.31	11.23	10.46	0.07	Olmsted (1962)
SL-1 1	433107112492201	3	0.5	19.5	2.47	0.37	0.45	0.05	0.03	73.31	11.1	15.58	0.03	Olmsted (1962)
SPERT 1	433252112520301	1.2	0.1	17.3	2.59	0.4	0.45	0.02	0.01	74.82	11.43	13.75	0.09	Olmsted (1962)
SPERT 2	433247112515201	2	0.4	12.4	2.62	0.48	0.28	0.03	0.02	76.31	13.94	9.76	0.01	Olmsted (1962)
TAN Exploration	435038112453401	2.7	0.29	22.9	2.28	0.5	0.56	0.04	0.02	66.96	14.81	18.23	0.07	Busenberg and others (2000, table 7)
USGS 1	432700112470801	3.8	0.57	17.4	2.59	0.27	0.37	0.06	0.03	78.04	8.16	13.8	-0.11	Busenberg and others (2000, table 7)
		1.8	0.7	12.5	2.46	0.29	0.28	0.03	0.04	79.36	9.41	11.23	-0.01	Olmsted (1962)
		1.2	0.6	10.8	2.43	0.29	0.25	0.02	0.03	80.27	9.65	10.09	-0.04	Olmsted (1962)
USGS 2	433320112432301	5.5	0.57	23.1	2.72	0.29	0.48	0.09	0.03	75.31	8.13	16.56	-0.11	Busenberg and others (2000, table 7)
		1.6	0.7	11.1	2.44	0.29	0.25	0.03	0.04	80.21	9.58	10.21	0	Olmsted (1962)
		1.5	0.7	11.2	2.41	0.25	0.25	0.02	0.04	81.01	8.4	10.59	0.06	Olmsted (1962)
USGS 3A	433732112335401	2.3	0.3	13.6	2.39	0.19	0.31	0.04	0.02	81.24	6.43	12.33	0.06	Olmsted (1962)
		1.5	0.9	10.4	2.44	0.19	0.23	0.02	0.05	83.38	6.47	10.15	0.02	Olmsted (1962)
USGS 4	434657112282201	13.4	0.18	56	5.62	0.71	1.2	0.22	0.01	72.51	9.16	18.34	0.06	Busenberg and others (2000, table 7)
		21.7	0.21	62.3	5.57	0.66	1.14	0.35	0.01	72.02	8.58	19.39	-0.34	Busenberg and others (2000, table 7)
		32	0.3	206.3	3.03	1.23	4.91	0.52	0.02	31.26	12.66	56.08	0.3	Olmsted (1962)
		29	0.3	189.3	3.05	1.19	4.51	0.47	0.02	33.02	12.86	54.12	0.13	Olmsted (1962)

Appendix D. Cation and anion data for selected wells at the Idaho National Laboratory, Idaho.—Continued

[Local name: is local well identifier used in this study. Concentrations are in milligrams per liter. Site identifier: is the unique numerical identifiers used to access well data (http://waterdata.usgs.gov/nwis). Wells with compositional and temporal variations in water type as noted by Olmsted (1962) are indicated in bold. Water type: is based on criteria used by Olmsted (1962). Abbreviations: Ca, calcium; Mg, magnesium; Sr, strontium; Na, sodium; K, potassium; CO_3, carbonate; HCO_3, bicarbonate; SO_4, sulfate; Cl, chloride; NO_3, nitrate; F, fluoride; mEq, milliequivalents; %, percent; –, not sampled]

Local name	Site identifier	NO_3	F	$Cl+NO_3+F$	CO_3+HCO_3 (mEq)	SO_4 (mEq)	Cl (mEq)	NO_3 (mEq)	F (mEq)	CO_3+HCO_3 % (mEq)	SO_4 % (mEq)	$Cl+NO_3+F$ % (mEq)	Cation anion balance	Reference
USGS 5	**433343112493801**	2	0.21	11.6	2.8	0.39	0.27	0.03	0.01	50.06	11.12	8.81	-0.12	Busenberg and others (2000, table 7)
		12	0.3	11.5	1.21	0.48	0.28	0.02	0.02	50.37	23.84	15.79	0.4	Olmsted (1962)
		13	0.2	11	2.82	0.42	0.27	0.02	0.01	79.75	11.78	8.47	-0.06	Olmsted (1962)
USGS 6	**434031112453701**	12	0.23	10.8	2.39	0.38	0.27	0.02	0.01	78.04	12.29	9.67	-0.13	Busenberg and others (2000, table 7)
		5.3	0.2	143.5	1.67	0.77	3.89	0.09	0.01	26	11.98	62.02	0.1	Olmsted (1962)
		14	0.4	11.3	2.46	0.42	0.27	0.02	0.02	77.15	13.07	9.78	-0.02	Olmsted (1962)
USGS 7	**434915112443901**	17	1.3	12.1	2.33	0.34	0.26	0.03	0.07	77.19	11.12	11.69	0	Busenberg and others (2000, table 7)
		5.5	0.4	15.9	2.74	0.48	0.28	0.09	0.02	75.87	13.27	10.86	0.18	Olmsted (1962)
		5	0.3	15.3	2.84	0.42	0.28	0.08	0.02	78.1	11.47	10.43	0.08	Olmsted (1962)
		4.6	0.2	12.8	2.79	0.42	0.23	0.07	0.01	79.31	11.85	8.83	0.05	Olmsted (1962)
		0.5	1.6	11.6	2.43	0.44	0.27	0.01	0.08	75.26	13.57	11.18	0.13	Olmsted (1962)
		0.6	–	10.6	2.41	0.42	0.28	0.01	0	77.28	13.36	9.36	0.01	Olmsted (1962)
		0.1	1.5	11.6	1.93	0.03	0.28	0	0.08	83.15	1.25	15.59	0.19	Olmsted (1962)
		0.1	1.6	11.7	2.33	0.15	0.28	0	0.08	81.68	5.41	12.91	0.12	Olmsted (1962)
		0.1	1.6	11.7	2.36	0.17	0.28	0	0.08	81.42	5.89	12.69	0.09	Olmsted (1962)
		0.1	1.6	11.7	2.39	0.16	0.28	0	0.08	82	5.39	12.61	0.06	Olmsted (1962)
USGS 8	**433121113115801**	3.8	0.2	12.4	3.29	0.34	0.24	0.06	0.01	83.46	8.72	7.82	-0.03	Busenberg and others (2000, table 7)
		13	0.3	11.6	2.95	0.5	0.28	0.02	0.02	73.28	13.26	8.46	-0.02	Olmsted (1962)
USGS 9	**432740113044501**	2.9	0.18	22.5	2.75	0.56	0.55	0.05	0.01	70.29	14.3	15.41	-0.09	Busenberg and others (2000, table 7)
		2.3	0.2	29.1	2.82	0.44	0.73	0.05	0.01	69.69	10.76	19.55	-0.11	Busenberg and others (2000, table 7)
		0.7	0.3	11	2.66	0.46	0.28	0.01	0.02	77.58	13.38	9.03	0.05	Olmsted (1962)
USGS 11	**432336113064201**	2.3	0.19	14.8	2.84	0.48	0.33	0.05	0.01	76.54	12.98	10.47	-0.07	Busenberg and others (2000, table 7)
		2.7	0.22	14.7	2.9	0.46	0.33	0.04	0.01	77.42	12.23	10.35	-0.12	Busenberg and others (2000, table 7)
		0.3	0.3	9.9	2.66	0.44	0.25	0.01	0.02	78.8	12.98	8.22	-0.04	Olmsted (1962)
		0.3	0.1	8.3	2.59	0.5	0.21	0.01	0.01	78.09	15.07	6.84	0.07	Olmsted (1962)
USGS 12	**434126112550701**	10.1	0.13	47.8	4.28	0.77	1.06	0.16	0.01	68.13	12.27	19.6	-0.06	Busenberg and others (2000, table 7)
		4.4	0	48.4	3.62	0.83	1.24	0.07	0	62.81	14.44	22.75	0.01	Olmsted (1962)
		4.1	0.2	44.3	3.57	0.79	1.13	0.07	0.01	64.16	14.21	21.64	0.11	Olmsted (1962)
USGS 13	**432737113143902**	2.2	0.2	57.4	2.47	0.6	1.55	0.04	0.01	52.93	12.91	34.16	0.03	Olmsted (1962)
USGS 14	**432019112563201**	4.3	0.79	26.6	2.75	0.45	0.59	0.08	0.04	70.38	11.44	18.18	-0.07	Olmsted (1962)

Appendix D. Cation and anion data for selected wells at the Idaho National Laboratory, Idaho.—Continued

[Local name: is local well identifier used in this study. Concentrations are in milligrams per liter. Site identifier: is the unique numerical identifiers used to access well data (http://waterdata.usgs.gov/nwis). Wells with compositional and temporal variations in water type as noted by Olmsted (1962) are indicated in bold. Water type: is based on criteria used by Olmsted (1962). Abbreviations: Ca, calcium; Mg, magnesium; Sr, strontium; Na, sodium; K, potassium; CO_3, carbonate; HCO_3, bicarbonate; SO_4, sulfate; Cl, chloride; NO_3, nitrate; F, fluoride; mEq, milliequivalents; %, percent; –, not sampled]

Local name	Site identifier	NO_3	F	Cl+NO_3 +F	CO_3+ HCO_3 (mEq)	SO_4 (mEq)	Cl (mEq)	NO_3 (mEq)	F (mEq)	CO_3+ HCO_3 % (mEq)	SO_4 % (mEq)	Cl+NO_3 +F % (mEq)	Cation anion balance	Reference
USGS 15	43423411255701	1.8	0.11	8.9	2.74	0.39	0.2	0.03	0.01	81 59	11.48	6.92	-0.04	Busenberg and others (2000, table 7)
		2.3	0.15	12.5	2.8	0.41	0.28	0.04	0.01	79.08	11.69	9.23	-0.05	Busenberg and others (2000, table 7)
		0.4	0.7	18.1	2.52	0.07	0.48	0.01	0.04	80.85	2.4	16.75	0.58	Olmsted (1962)
		1.2	0.1	8.3	2.67	0.4	0.2	0.02	0.01	81.22	12.03	6.75	0.02	Olmsted (1962)
USGS 17	43393711251540 1	1.5	0.21	6.6	2.47	0.4	0.14	0.02	0.01	81 25	13.06	5.69	0.01	Busenberg and others (2000, table 7)
		0.1	0.2	7.1	2.34	0.33	0.19	0	0.01	81 36	11.56	7.08	0	Olmsted (1962)
		0.5	0.2	7.7	1.64	0.44	0.2	0.01	0.01	71 5	19.08	9.42	0.01	Olmsted (1962)
		0.5	0.2	7.3	1.33	0.44	0.19	0.01	0.01	67.4	22.2	10.4	0.33	Olmsted (1962)
USGS 18	43454011244090 1	1.9	0.3	12.4	2.75	0.51	0.29	0.03	0.02	76.44	14.28	9.28	0.05	Busenberg and others (2000, table 7)
		1.4	0.3	10.9	2.72	0.56	0.26	0.02	0.02	75 98	15.7	8.32	0.06	Olmsted (1962)
USGS 19	43442611257570 1	3.8	0.14	12.8	3.36	0.46	0.25	0.06	0.01	81.08	11.21	7.71	-0.09	Busenberg and others (2000, table 7)
		3.7	0.21	13.8	3.23	0.43	0.28	0.06	0.01	80 56	10.7	8.73	0.08	Busenberg and others (2000, table 7)
USGS 20	43325311254590 1	3.1	0.1	17.2	3.05	0.54	0.39	0.05	0.01	75.46	13.4	11.14	0.07	Olmsted (1962)
		1.2	0.3	54.5	2.29	0.52	1.49	0.02	0.02	52.81	11.98	35.21	-0.04	Olmsted (1962)
USGS 21	43430711238260 1	0.8	0.1	12.9	2.36	0.35	0.34	0.01	0.01	76.86	11.53	11.61	-0.02	Olmsted (1962)
USGS 22	43342211303170 1	2.8	0.3	17.1	2.29	0.52	0.39	0.05	0.02	70 15	15.91	13.94	0.03	Olmsted (1962)
		1.8	0.17	68.5	1.43	0.44	1.88	0.03	0.01	37.75	11.58	50.67	-0.12	Busenberg and others (2000, table 7)
USGS 23	43405511259590 1	4.1	0.2	51.3	1.44	0.46	1.33	0.07	0.01	43.67	13.87	42.46	0.03	Olmsted (1962)
		2.5	0.21	12.6	2.98	0.37	0.28	0.04	0.01	81.06	9.96	8.98	-0.07	Busenberg and others (2000, table 7)
USGS 24	43505311242080 1	2.3	0.3	13.6	2.88	0.42	0.31	0.04	0.02	78.72	11.37	9.91	0.13	Olmsted (1962)
USGS 25	43533911244460 1	2.2	0.2	9.8	3.1	0.56	0.21	0.04	0.01	79 13	14.36	6.51	0.05	Olmsted (1962)
		1.6	0.1	8.2	3.06	0.6	0.18	0.03	0.01	78 93	15.55	5.52	0.07	Olmsted (1962)
USGS 26	43521211239400 1	3.4	0.43	17.1	2.98	0.6	0.38	0.05	0.02	74	14.77	11.23	-0.09	Busenberg and others (2000, table 7)
USGS 27	43485111232180 1	1.7	0.5	13.2	2.93	0.62	0.31	0.03	0.03	74.79	15.93	9.28	0	Olmsted (1962)
		10.6	0.58	73.1	2.79	0.8	1.75	0.17	0.03	50 33	14.48	35.18	-0.06	Busenberg and others (2000, table 7)
USGS 28	43460011236010 1	1.3	0.5	10.8	2.69	0.52	0.25	0.02	0.03	76 59	14.83	8.58	0.03	Olmsted (1962)

Appendix D. Cation and anion data for selected wells at the Idaho National Laboratory, Idaho.—Continued

[**Local name:** is local well identifier used in this study. Concentrations are in milligrams per liter. **Site identifier:** is the unique numerical identifiers used to access well data (http://waterdata.usgs.gov/nwis). Wells with compositional and temporal variations in water type as noted by Olmsted (1962) are indicated in bold. **Water type:** is based on criteria used by Olmsted (1962). **Abbreviations:** Ca, calcium; Mg, magnesium; Sr, strontium; Na, sodium; K, potassium; CO_3, carbonate; HCO_3, bicarbonate; SO_4, sulfate; Cl, chloride; NO_3, nitrate; F, fluoride; mEq, milliequivalents; %, percent; –, not sampled]

Local name	Site identifier	NO_3	F	$Cl+NO_3+F$	CO_3+HCO_3 (mEq)	SO_4 (mEq)	Cl (mEq)	NO_3 (mEq)	F (mEq)	CO_3+HCO_3 % (mEq)	SO_4 % (mEq)	$Cl+NO_3+F$ % (mEq)	Cation anion balance	Reference
USGS 29	434407112285101	9	0.43	35.4	3.15	0.35	0.73	0.15	0.02	71.59	7.91	20.5	0.1	Busenberg and others (2000, table 7)
		8.7	0.42	36.3	3.2	0.33	0.77	0.14	0.02	71.68	7.47	20.85	–	Busenberg and others (2000, table 7)
		2	0.5	34.5	1.64	0.4	0.9	0.03	0.03	54.71	13.21	32.09	0.11	Olmsted (1962)
USGS 30		0.9	0.7	15.6	2.1	0.58	0.39	0.01	0.04	67.09	18.64	14.27	-0.07	Olmsted (1962)
		1	0.6	15.6	2.82	0.56	0.39	0.02	0.03	73.72	14.7	11.58	-0.07	Olmsted (1962)
USGS 31	434625112342101	3.5	0.42	25.2	2.77	0.6	0.6	0.06	0.02	68.47	14.77	16.76	0.02	Busenberg and others (2000, table 7)
		3.5	0.41	26.5	2.82	0.58	0.63	0.06	0.02	68.59	14.03	17.38	-0.08	Busenberg and others (2000, table 7)
		1.3	0.5	11.8	2.77	0.5	0.28	0.02	0.03	76.96	13.89	9.15	-0.04	Olmsted (1962)
USGS 32	434444112322101	6.2	0.38	48.6	2.67	0.81	1.18	0.1	0.02	55.77	17	27.24	0.1	Busenberg and others (2000, table 7)
USGS 33	434314112322901	2.5	0.6	21.2	2.69	0.56	0.51	0.04	0.03	70.16	14.67	15.17	0.09	Olmsted (1962)
		9.9	0.6	79.5	2.69	0.67	1.95	0.16	0.03	48.94	12.13	38.92	0.18	Olmsted (1962)
USGS 34	433334112565501	–	–	–	–	–	–	–	–	–	–	–	–	Olmsted (1962)
USGS 36	433330112565201	8	0.22	42.1	3.26	0.58	0.96	0.13	0.01	66.09	11.69	22.22	0.17	Busenberg and others (2000, table 7)
USGS 40	433411112561101	5.1	0.3	86.4	2.36	0.5	2.28	0.08	0.02	45.02	9.53	45.45	-0.03	Olmsted (1962)
		14	0.3	140.3	3.08	0.71	3.55	0.23	0.02	40.62	9.33	50.04	-0.02	Olmsted (1962)
USGS 41	433491112561301	12	0.3	138.3	3.1	0.71	3.55	0.19	0.01	40.93	9.35	49.72	-0.12	Olmsted (1962)
USGS 42	433441112561301	5.2	0.2	59.4	3.11	0.52	1.52	0.08	0.01	59.29	9.91	30.8	0.11	Olmsted (1962)
		4.2	0.3	44.5	3.05	0.52	1.13	0.07	0.02	63.76	10.89	25.35	0.09	Olmsted (1962)
		5.9	0.3	62.2	3.21	0.56	1.58	0.1	0.02	58.78	10.29	30.93	0.05	Olmsted (1962)
		5.	0.3	53.4	3.06	0.56	1.35	0.08	0.02	60.34	11.07	28.59	0.15	Olmsted (1962)
USGS 43	433415112561501	1.	0.5	9.6	2.11	0.44	0.23	0.02	0.03	74.94	15.5	9.56	0.05	Olmsted (1962)
		7.7	0.3	85	3.05	0.6	2.17	0.12	0.02	51.11	10.12	38.76	0.16	Olmsted (1962)
		7.9	0.3	92.2	3.18	0.58	2.37	0.13	0.02	50.67	9.29	40.04	0.02	Olmsted (1962)
USGS 44	433499112562101	0.5	0.4	7.4	2.64	0.42	0.18	0.01	0.02	80.75	12.74	6.5	-0.07	Olmsted (1962)
		1.8	0.2	12.8	3.08	0.46	0.31	0.03	0.01	76.29	11.79	8.92	0.14	Olmsted (1962)
		1.9	0.3	12.2	2.85	0.44	0.28	0.03	0.02	75.83	12.09	9.08	0.06	Olmsted (1962)
USGS 45	433402112561801	–	–	–	–	–	–	–	–	–	–	–	4.26	Olmsted (1962)
USGS 46	433407112561501	4.	0.2	50.3	3.11	0.54	1.3	0.07	0.01	61.91	10.76	27.32	0.12	Olmsted (1962)
		2.2	0.3	23.5	3.15	0.44	0.59	0.04	0.02	74.43	10.34	15.22	0.06	Olmsted (1962)
		4.5	0.3	52.6	3.18	0.48	1.35	0.07	0.02	62.37	9.39	28.23	0.08	Olmsted (1962)
		3.8	0.3	44.1	3.18	0.52	1.13	0.06	0.02	64.82	10.61	24.57	0	Olmsted (1962)
USGS 47	433407112560301	9	0.1	163.1	3.16	0.81	4.34	0.15	0.01	37.35	9.59	53.06	-0.12	Olmsted (1962)
USGS 48	433401112560301	8.6	0.2	128.8	3.1	0.67	3.38	0.14	0.01	42.44	9.13	48.43	0.03	Olmsted (1962)

Appendix D. Cation and anion data for selected wells at the Idaho National Laboratory, Idaho.—Continued

[Local name: is local well identifier used in this study. Concentrations are in milligrams per liter. Site identifier: is the unique numerical identifiers used to access well data (http://waterdata.usgs.gov/nwis). Wells with compositional and temporal variations in water type as noted by Olmsted (1962) are indicated in bold. Water type: is based on criteria used by Olmsted (1962). Abbreviations: Ca, calcium; Mg, magnesium; Sr, strontium; Na, sodium; K, potassium; CO_3, carbonate; HCO_3, bicarbonate; SO_4, sulfate; Cl, chloride; NO_3, nitrate; F, fluoride; mEq, milliequivalents; %, percent; –, not sampled]

Local name	Site identifier	NO_3	F	$Cl+NO_3$ $+F$	CO_3+ HCO_3 (mEq)	SO_4 (mEq)	Cl (mEq)	NO_3 (mEq)	F (mEq)	CO_3+ HCO_3 % (mEq)	SO_4 % (mEq)	$Cl+NO_3$ $+F$ % (mEq)	Cation anion balance	Reference
USGS 49	433403112555401	8.5	0.3	76.8	3.28	0.62	1.92	0.14	0.02	54.87	10.46	34.67	0.04	Olmsted (1962)
		5.6	0.3	78.9	3.11	0.6	2.06	0.09	0.02	52.93	10.26	36.8	-0.02	Olmsted (1962)
		6.4	0.3	84.7	3.15	0.6	2.2	0.1	0.02	51.84	9.95	38.21	-0.02	Olmsted (1962)
		6.4	0.3	82.7	3.15	0.56	2.14	0.1	0.02	52.69	9.41	37.89	-0.06	Olmsted (1962)
USGS 51	433350112560601	2.6	0.2	42.8	2.77	0.48	1.13	0.04	0.01	62.53	10.81	26.66	0.07	Olmsted (1962)
USGS 55	433508112573001	11	0	57	2.87	9.66	1.3	0.18	0	20.48	68.99	10.53	0.19	Olmsted (1962)
USGS 57	433344112562601	8.4	0.23	–	–	–	–	–	–	–	–	–	–	Busenberg and others (2000, table 7)
		11.2	0.22	–	–	–	–	–	–	–	–	–	–	Busenberg and others (2000, table 7)
USGS 65	433447112574501	3.1	0.2	19.3	2.44	1.25	0.45	0.05	0.01	58.1	29.72	12.18	0.06	Olmsted (1962)
USGS 66	433436112564801	3.8	0.8	20.6	3.18	0.92	0.45	0.06	0.04	68.37	19.7	11.93	-0.02	Olmsted (1962)
USGS 77	433315112560301	18.8	0.22	–	–	–	–	–	–	–	–	–	–	Busenberg and others (2000, table 7)
USGS 82	433401112551001	2.4	0.21	20.6	2.49	0.44	0.51	0.04	0.01	71.46	12.54	15.99	-0.07	Busenberg and others (2000, table 7)
USGS 83	433023112561501	2.9	0.24	13.9	2.02	0.42	0.3	0.05	0.01	72.04	14.96	13.01	-0.08	Busenberg and others (2000, table 7)
USGS 86	432935113080001	6.3	0.16	26.1	2.16	0.47	0.55	0.1	0.01	65.58	14.33	20.09	-0.06	Busenberg and others (2000, table 7)
USGS 89	433005113032801	8	0.33	47.1	1.69	0.73	1.09	0.13	0.02	46.18	19.88	33.94	-0.14	Busenberg and others (2000, table 7)
USGS 97	433807112551501	9.6	0.2	47.8	4.41	0.75	1.07	0.15	0.01	68.96	11.69	19.35	-0.02	Busenberg and others (2000, table 7)
USGS 98	433657112563601	4.8	0.12	20.1	3.43	0.45	0.43	0.08	0.01	78.03	10.29	11.67	0.05	Busenberg and others (2000, table 7)
USGS 99	433705112552101	6.7	0.15	29.1	4.05	0.56	0.63	0.11	0.01	75.63	10.5	13.87	0.07	Busenberg and others (2000, table 7)
USGS 100	433503112400701	6.5	0.57	24.8	2.69	0.44	0.5	0.1	0.03	71.5	11.63	16.87	-0.06	Busenberg and others (2000, table 7)
		6.1	0.6	23.1	2.77	0.31	0.46	0.1	0.03	75.46	8.4	16.14	-0.13	Busenberg and others (2000, table 7)
USGS 101	433255112381801	3.6	0.77	12.9	2.38	0.18	0.24	0.06	0.04	82	6.32	11.68	-0.07	Busenberg and others (2000, table 7)
		3.6	0.78	12.9	2.43	0.19	0.24	0.06	0.04	82 17	6.35	11.48	-0.12	Busenberg and others (2000, table 7)
USGS 102	433853112551601	9.1	0.13	43.2	4.33	0.74	0.96	0.15	0.01	70.03	11.96	18.01	0.06	Busenberg and others (2000, table 7)

Appendix D. Cation and anion data for selected wells at the Idaho National Laboratory, Idaho.—Continued

[**Local name:** is local well identifier used in this study. Concentrations are in milligrams per liter. **Site identifier:** is the unique numerical identifiers used to access well data (http://waterdata.usgs.gov/nwis). Wells with compositional and temporal variations in water type as noted by Olmsted (1962) are indicated in bold. **Water type:** is based on criteria used by Olmsted (1962). **Abbreviations: Ca,** calcium; Mg, magnesium; Sr, strontium; Na, sodium; K, potassium; CO_3, carbonate; HCO_3, bicarbonate; SO_4, sulfate; Cl, chloride; NO_3, nitrate; F, fluoride; mEq, milliequivalents; %, percent; –, not sampled]

Local name	Site identifier	NO_3	F	Cl+NO_3+F	CO_3+HCO_3 (mEq)	SO_4 (mEq)	Cl (mEq)	NO_3 (mEq)	F (mEq)	CO_3+HCO_3 % (mEq)	SO_4 % (mEq)	Cl+NO_3+F % (mEq)	Cation anion balance	Reference
USGS 103	432714112560701	3.3	0.31	20.7	2.74	0.5	0.48	0.05	0.02	72.16	13.29	14.55	-0.13	Busenberg and others (2000, table 7)
		3.3	0.32	19.9	2.74	0.48	0.46	0.05	0.02	73.03	12.83	14.14	-0.06	Busenberg and others (2000, table 7)
	432714112560723	0.1	0.38	18.9	2.02	0.48	0.52	0	0.02	66.35	15.83	17.82	-0.04	U.S. Geological Survey (2011)
	432714112560720	0.4	0.32	15.3	2.66	0.48	0.4	0.01	0.02	74.53	13.38	12.09	-0.09	U.S. Geological Survey (2011)
	432714112560716	0.5	0.22	10.9	2.59	0.41	0.29	0.01	0.01	73.36	12.35	9.3	-0.17	U.S. Geological Survey (2011)
	432714112560712	0.5	0.22	13.5	2.82	0.45	0.36	0.01	0.01	77.39	12.23	10.38	-0.16	U.S. Geological Survey (2011)
	432714112560708	0.8	0.21	15.3	2.88	0.46	0.4	0.01	0.01	76.4	12.3	11.3	-0.26	U.S. Geological Survey (2011)
	432714112560704	0.8	0.21	15.1	2.88	0.47	0.4	0.01	0.01	76.47	12.36	11.17	-0.1	U.S. Geological Survey (2011)
	432714112560702	0.8	0.21	15.3	2.88	0.47	0.4	0.01	0.01	75.31	12.39	11.29	-0.15	U.S. Geological Survey (2011)
USGS 104	432856112560801	3.2	0.19	15.2	2.52	0.41	0.33	0.05	0.01	75.83	12.32	11.85	-0.04	Busenberg and others (2000, table 7)
		3.2	0.2	16	2.56	0.4	0.36	0.05	0.01	75.73	11.9	12.37	-0.12	Busenberg and others (2000, table 7)
USGS 105	432703113001801	3	0.19	16.8	2.95	0.54	0.38	0.05	0.01	75	13.76	11.24	-0.02	Busenberg and others (2000, table 7)
	432703113001818	–	–	12.8	–	–	–	–	–	–	–	–	–	U.S. Geological Survey (2011)
	432703113001815	–	–	12.9	–	–	–	–	–	–	–	–	–	U.S. Geological Survey (2011)
	432703113001811	–	–	13.9	–	–	–	–	–	–	–	–	–	U.S. Geological Survey (2011)
	432703113001807	–	–	13.6	–	–	–	–	–	–	–	–	–	U.S. Geological Survey (2011)
	432703113001803	–	–	12	–	–	–	–	–	–	–	–	–	U.S. Geological Survey (2011)
USGS 107	432942112532801	4.6	0.34	26.2	2.88	0.53	0.6	0.07	0.02	73.28	12.84	16.88	–	Busenberg and others (2000, table 7)
USGS 108	432659112582601	2.9	0.24	17.1	2.7	0.47	0.39	0.05	0.01	74.6	12.87	12.53	–	Busenberg and others (2000, table 7)

Appendix D.   Cation and anion data for selected wells at the Idaho National Laboratory, Idaho.—Continued

[**Local name:** is local well identifier used in this study. Concentrations are in milligrams per liter. **Site identifier:** is the unique numerical identifiers used to access well data (http://waterdata.usgs.gov/nwis). Wells with compositional and temporal variations in water type as noted by Olmsted (1962) are indicated in bold. **Water type:** is based on criteria used by Olmsted (1962). **Abbreviations:** Ca, calcium; Mg, magnesium; Sr, strontium; Na, sodium; K, potassium; CO_3, carbonate; HCO_3, bicarbonate; SO_4, sulfate; Cl, chloride; NO_3, nitrate; F, fluoride; mEq, milliequivalents; %, percent; –, not sampled]

Local name	Site identifier	NO_3	F	Cl+NO_3 +F	CO_3 + HCO_3 (mEq)	SO_4 (mEq)	Cl (mEq)	NO_3 (mEq)	F (mEq)	CO_3 + HCO_3 % (mEq)	SO_4 % (mEq)	Cl+NO_3 +F % (mEq)	Cation anion balance	Reference
USGS 109	432701113025601	2.9	0.2	22.5	2.88	0.56	0.55	0.05	0.01	71.24	13.83	14.93	–	Busenberg and others (2000, table 7)
		2.5	0.23	16.7	2.97	0.52	0.39	0.04	0.01	75.4	13.23	11.37	–	Busenberg and others (2000, table 7)
USGS 110A	432717112501502	5	0.45	24.5	2.84	0.37	0.54	0.08	0.02	73.64	9.73	16.63	–	Busenberg and others (2000, table 7)
USGS 112	433314112563001	14	0.26	165.3	2.84	0.6	4.26	0.23	0.01	35.72	7.61	56.67	–	Busenberg and others (2000, table 7)
USGS 113	433314112561801	10.6	0.15	228.8	2.69	0.65	6.15	0.17	0.01	27.81	6.72	65.47	-0.29	Busenberg and others (2000, table 7)
USGS 115	433320112554101	5.6	0.23	43.8	2.39	0.44	1.07	0.09	0.01	59.69	11.01	29.29	-0.11	Busenberg and others (2000, table 7)
USGS 116	433331112553201	13.2	0.3	102.8	2	0.71	2.52	0.21	0.02	36.63	13.04	50.33	-0.13	Busenberg and others (2000, table 7)
USGS 117	432955113025901	2.6	0.22	16.5	1.98	0.36	0.39	0.04	0.01	71.36	12.81	15.83	-0.09	Busenberg and others (2000, table 7)
USGS 120	432919113031501	3.4	0.26	25.4	3.05	0.79	0.61	0.05	0.01	67.44	17.5	15.06	-0.1	Busenberg and others (2000, table 7)
USGS 124	432307112583101	3.3	0.31	17.9	2.82	0.47	0.4	0.05	0.02	75.01	12.41	12.58	-0.04	Busenberg and others (2000, table 7)
		3.2	0.3	18.3	2.88	0.45	0.42	0.05	0.02	75 57	11.73	12.7	-0.11	Busenberg and others (2000, table 7)
USGS 125	432602113052801	2.8	0.21	17.9	2.92	0.54	0.42	0.05	0.01	74.21	13.67	12.12	0	Busenberg and others (2000, table 7)
		2.5	0.23	17.5	2.97	0.51	0.42	0.04	0.01	75.17	12.93	11.91	-0.11	Busenberg and others (2000, table 7)
USGS 126A	435529112471301	–	–	8.7	–	–	–	–	–	–	5	–	–	Swanson and others (2003)
USGS 126B	435529112471401	–	–	8.9	–	–	–	–	–	–	6	–	–	Swanson and others (2003)
USGS 127	433058112572201	–	–	15.7	–	–	–	–	–	–	–	–	–	U.S. Geological Survey (2011)
USGS 128	433250112565601	–	–	55.7	–	–	–	–	–	–	–	–	–	U.S. Geological Survey (2011)

Appendix D. Cation and anion data for selected wells at the Idaho National Laboratory, Idaho.—Continued

[**Local name:** is local well 1 identifier used in this study. Concentrations are in milligrams per liter. **Site identifier:** is the unique numerical identifiers used to access well data (http://waterdata.usgs.gov/nwis). Wells with compositional and temporal variations in water type as noted by Olmsted (1962) are indicated in bold. **Water type:** is based on criteria used by Olmsted (1962). **Abbreviations:** Ca, calcium; Mg, magnesium; Sr, strontium; Na, sodium; K, potassium; CO₃, carbonate; HCO₃, bicarbonate; SO₄, sulfate; Cl, chloride; NO₃, nitrate; F, fluoride; mEq, milliequivalents; %, percent; –, not sampled]

Local name	Site identifier	NO₃	F	Cl+NO₃+F	CO₃+HCO₃ (mEq)	SO₄ (mEq)	Cl (mEq)	NO₃ (mEq)	F (mEq)	CO₃+HCO₃ % (mEq)	SO₄ % (mEq)	Cl+NO₃+F % (mEq)	Cation anion balance	Reference
USGS 132	432906113025022	1.3	0.28	36.5	2.9	0.85	0.99	0.02	0.01	60.78	17.8	21.42	-0.26	U.S. Geological Survey (2011)
	432906113025018	0.7	0.26	12.2	2.95	0.54	0.32	0.01	0.01	77.01	14.08	8.91	-0.11	U.S. Geological Survey (2011)
	432906113025014	0.7	0.25	11.2	2.92	0.52	0.29	0.01	0.01	77.83	13.78	8.39	-0.13	U.S. Geological Survey (2011)
	432906113025010	0.7	0.24	11.1	2.92	0.51	0.29	0.01	0.01	77.95	13.74	8.31	-0.15	U.S. Geological Survey (2011)
	432906113025006	0.7	0.25	11.2	2.92	0.52	0.29	0.01	0.01	77.83	13.78	8.39	-0.09	U.S. Geological Survey (2011)
	432906113025001	0.8	0.25	11.7	3.03	0.53	0.3	0.01	0.01	77.99	13.6	8.41	-0.16	U.S. Geological Survey (2011)
USGS 133	433635112554312	0.4	0.23	13.1	2.75	0.43	0.35	0.01	0.01	77.43	12.12	10.45	0.19	U.S. Geological Survey (2011)
	433635112554308	0.9	0.2	14	2.85	0.44	0.36	0.01	0.01	77.53	11.89	10.58	-0.21	U.S. Geological Survey (2011)
	433635112554305		0.19	12.7	2.9	0.43	0.35	0	0.01	78.61	11.57	9.83	-0.2	U.S. Geological Survey (2011)
	433635112554301	1.2	0.13	15.2	3.75	0.46	0.39	0.02	0.01	81.03	9.93	9.03	-0.22	U.S. Geological Survey (2011)
USGS 134	433611112595819	–	–	9.8	–	–	–	–	–	–	–	–	–	U.S. Geological Survey (2011)
	433611112595815	–	–	10.5	–	–	–	–	–	–	–	–	–	U.S. Geological Survey (2011)
	433611112595811	–	–	7.9	–	–	–	–	–	–	–	–	–	U.S. Geological Survey (2011)
	433611112595807	–	–	10.6	–	–	–	–	–	–	–	–	–	U.S. Geological Survey (2011)
	433611112595804	0.4	0.13	8	2.59	0.4	0.21	0.01	0.01	80.61	12.38	7.01	-0.27	U.S. Geological Survey (2011)
USGS 135	432753113093613	–	–	8.5	–	–	–	–	–	–	–	–	–	U.S. Geological Survey (2011)
	432753113093609	–	–	8.1	–	–	–	–	–	–	–	–	–	U.S. Geological Survey (2011)
	432753113093605	–	–	8.9	–	–	–	–	–	–	–	–	–	U.S. Geological Survey (2011)
	432753113093601	–	–	8.4	–	–	–	–	–	–	–	–	–	U.S. Geological Survey (2011)

Appendix D.   Cation and anion data for selected wells at the Idaho National Laboratory, Idaho.—Continued

[**Local name:** is local well identifier used in this study. Concentrations are in milligrams per liter. **Site identifier:** is the unique numerical identifiers used to access well data (http://waterdata.usgs.gov/nwis). Wells with compositional and temporal variations in water type as noted by Olmsted (1962) are indicated in bold. **Water type:** is based on criteria used by Olmsted (1962). **Abbreviations:** Ca, calcium; Mg, magnesium; Sr, strontium; Na, sodium; K, potassium; CO_3, carbonate; HCO_3, bicarbonate; SO_4, sulfate; Cl, chloride; NO_3, nitrate; F, fluoride; mEq, milliequivalents; %, percent; –, not sampled]

Local name	Site identifier	NO_3	F	$Cl+NO_3$ +F	CO_3 + HCO_3 (mEq)	SO_4 (mEq)	Cl (mEq)	NO_3 (mEq)	F (mEq)	CO_3 + HCO_3 % (mEq)	SO_4 % (mEq)	$Cl+NO_3$ +F % (mEq)	Cation anion balance	Reference
MIDDLE 2050A	433409112570515	–	–	16.7	–	–	–	–	–	–	–	–	–	U.S. Geological Survey (2011)
	433409112570512	–	–	11.8	–	–	–	–	–	–	–	–	–	U.S. Geological Survey (2011)
	433409112570509	–	–	12	–	–	–	–	–	–	–	–	–	U.S. Geological Survey (2011)
	433409112570506	–	–	11.4	–	–	–	–	–	–	–	–	–	U.S. Geological Survey (2011)
	433409112570503	–	–	14.9	–	–	–	–	–	–	–	–	–	U.S. Geological Survey (2011)
MIDDLE 2051	433217113004912	–	–	6.3	–	–	–	–	–	–	–	–	–	U.S. Geological Survey (2011)
	433217113004909	–	–	11.7	–	–	–	–	–	–	–	–	–	U.S. Geological Survey (2011)
	433217113004906	–	–	12.2	–	–	–	–	–	–	–	–	–	U.S. Geological Survey (2011)
	433217113004903	–	–	13.2	–	–	–	–	–	–	–	–	–	U.S. Geological Survey (2011)
	433217113004901	–	–	13.1	–	–	–	–	–	–	–	–	–	U.S. Geological Survey (2011)

Appendix E. Concentrations of Lithium and Boron in Water from Sampling Sites Located at the Idaho National Laboratory and Vicinity, Idaho

[**Local name:** is the local well identifier used in this study. **Site identifier:** the unique numerical identifiers used to access well data within the National Water Information System (NWIS, http://waterdata.usgs.gov/nwis). **Lithium and boron:** concentrations in micrograms per liter. **Abbreviations:** µg/L, micrograms per liter; NS, not sampled; NA, not applicable; –, not available]

Local name	Site identifier	Date sampled	Lithium (µg/L)	Boron (µg/L)	Ratio of boron to lithium	Reference
ANP 6	435152112443101	06-15-95	3	26	9	Busenberg and others, 2000
		07-19-96	2.9	26	9	Busenberg and others, 2000
ANP 9	434856112400001	10-14-96	10.2	35	3	Busenberg and others, 2000
Arbor Test 1	433509112384801	04-21-95	24.7	44	2	Busenberg and others, 2000
		10-10-96	24.9	46	2	Busenberg and others, 2000
Arco City Well 4	433758113181701	05-13-97	1	11	11	Busenberg and others, 2000
Area 2	433223112470201	07-18-96	17.7	41	2	Busenberg and others, 2000
Atomic City	432638112484101	10-09-06	18	40	2	Busenberg and others, 2000
BFW	433042112535101	07-16-96	3.9	21	5	Busenberg and others, 2000
CFA 1	433204112562001	06-19-91	5	NS	NA	Liszewski and Mann, 1993
		07-16-96	2.5	21	8	Busenberg and others, 2000
CFA 2	433144112563501	07-16-96	3.6	27	8	Busenberg and others, 2000
EBR 1	433051113002601	06-19-91	4	NS	NA	Liszewski and Mann, 1993
		10-16-96	2.7	20	7	Busenberg and others, 2000
Engberson Well (ML-9)	435028112264501	05-14-97	14.4	36	3	Busenberg and others, 2000
Fire Station 2	433548112562301	06-19-91	4	NS	NA	Liszewski and Mann, 1993
		10-16-96	2	24	12	Busenberg and others, 2000
IET 1 Disposal	435153112420501	07-18-96	2.3	NS	NA	Busenberg and others, 2000
INEL-1 WS	433717112563501	06-12-95	2.8	19	7	Busenberg and others, 2000
Leo Rogers 1	432533112504901	07-17-96	16	40	3	Busenberg and others, 2000
Neville Well (ML-7)	435540112092901	05-14-97	25	58	2	Busenberg and others, 2000
NPR Test	433449112523101	06-20-91	4	NS	NA	Liszewski and Mann, 1993
		04-17-95	2	16	8	Busenberg and others, 2000
		10-10-96	2.2	NS	NA	Busenberg and others, 2000
Pancheri 6	435728113103701	05-13-97	1.5	16	11	Busenberg and others, 2000
Park Bell (08N 34E 17CCC7)	440058112293605	06-11-91	71	NS	NA	Liszewski and Mann, 1993
		05-21-97	73.5	84	1	Busenberg and others, 2000
PSTF Test	434941112454201	10-14-96	1.8	19	11	Busenberg and others, 2000
P&W 2	435419112453101	04-19-95	2.9	NS	NA	Busenberg and others, 2000
		10-15-96	2.9	18	6	Busenberg and others, 2000
RWMC M3S	433008113021801	07-22-96	2.4	18	8	Busenberg and others, 2000
RWMC M7S	433023113014801	07-22-96	2.2	17	8	Busenberg and others, 2000
Site 4	433617112542001	10-16-96	1.7	20	12	Busenberg and others, 2000
Site 9	433123112530101	06-25-91	4	NS	NA	Liszewski and Mann, 1993
		07-22-96	3.5	30	9	Busenberg and others, 2000

Appendix E.   Concentrations of lithium and boron in water from sampling sites located at the Idaho National Laboratory and vicinity, Idaho.—Continued

[**Local name:** is the local well identifier used in this study. **Site identifier:** the unique numerical identifiers used to access well data within the National Water Information System (NWIS, http://waterdata.usgs.gov/nwis). **Lithium and boron:** concentrations in micrograms per liter. **Abbreviations:** μg/L, micrograms per liter; NS, not sampled; NA, not applicable; –, not available]

Local name	Site identifier	Date sampled	Lithium (µg/L)	Boron (µg/L)	Ratio of boron to lithium	Reference
Site 14	434334112463101	06-13-91	13	NS	NA	Liszewski and Mann, 1993
		10-14-96	11.5	35	3	Busenberg and others, 2000
Site 17	434027112575701	06-18-91	5	NS	NA	Liszewski and Mann, 1993
		06-18-91	5	NS	NA	Liszewski and Mann, 1993
		06-16-95	2.4	26	11	Busenberg and others, 2000
Site 19	433522112582101	05-09-91	4	NS	NA	Liszewski and Mann, 1993
		07-16-96	2.5	25	10	Busenberg and others, 2000
TAN Exploration	435038112453401	10-14-96	2.5	20	8	Busenberg and others, 2000
USGS 1	432700112470801	05-30-91	22	NS	NA	Liszewski and Mann, 1993
		10-09-96	18	42	2	Busenberg and others, 2000
USGS 2	433320112432301	05-28-91	22	NS	NA	Liszewski and Mann, 1993
		05-28-91	22	NS	NA	Liszewski and Mann, 1993
		07-17-96	20.4	45	2	Busenberg and others, 2000
USGS 4	434657112282201	06-04-91	25	NS	NA	Liszewski and Mann, 1993
		06-04-91	27	NS	NA	Liszewski and Mann, 1993
		04-19-95	24.2	48	2	Busenberg and others, 2000
		10-15-96	23.7	NS	NA	Busenberg and others, 2000
USGS 5	433543112493801	10-10-96	2	19	10	Busenberg and others, 2000
USGS 6	434031112453701	07-18-96	7.3	25	3	Busenberg and others, 2000
USGS 7	434915112443901	05-20-91	27	NS	NA	Liszewski and Mann, 1993
		05-20-91	28	NS	NA	Liszewski and Mann, 1993
		10-14-96	25.9	57	2	Busenberg and others, 2000
USGS 8	433121113115801	05-31-91	6	NS	NA	Liszewski and Mann, 1993
		10-08-96	1.3	13	10	Busenberg and others, 2000
USGS 9	432740113044501	05-31-91	4	NS	NA	Liszewski and Mann, 1993
		04-20-95	3.2	22	7	Busenberg and others, 2000
		10-11-96	3.3	NS	NA	Busenberg and others, 2000
USGS 11	432336113064201	04-20-95	2.1	16	8	Busenberg and others, 2000
		10-09-95	2.1	NS	NA	Busenberg and others, 2001
USGS 12	434126112550701	06-14-95	2.7	33	12	Busenberg and others, 2000
USGS 14	432019112563201	10-09-96	24.3	36	1	Busenberg and others, 2000
USGS 15	434234112551701	06-14-95	2.1	18	9	Busenberg and others, 2000
		05-13-97	2.1	NS	NA	Busenberg and others, 2000
USGS 17	433937112515401	06-06-91	4	NS	NA	Liszewski and Mann, 1993
		06-13-95	1.4	13	9	Busenberg and others, 2000
USGS 18	434540112440901	07-19-96	5.2	33	6	Busenberg and others, 2000
USGS 19	434426112575701	05-21-91	5	NS	NA	Liszewski and Mann, 1993
		04-19-95	3.5	33	9	Busenberg and others, 2000
		10-15-96	3.5	NS	NA	Busenberg and others, 2000
USGS 22	433422113031701	06-13-95	3.7	33	9	Busenberg and others, 2000
		07-18-96	3.8	36	9	Busenberg and others, 2000

Appendix E.   Concentrations of lithium and boron in water from sampling sites located at the Idaho National Laboratory and vicinity, Idaho.—Continued

[**Local name:** is the local well identifier used in this study. **Site identifier:** the unique numerical identifiers used to access well data within the National Water Information System (NWIS, http://waterdata.usgs.gov/nwis). **Lithium and boron:** concentrations in micrograms per liter. **Abbreviations:** µg/L, micrograms per liter; NS, not sampled; NA, not applicable; –, not available]

Local name	Site identifier	Date sampled	Lithium (µg/L)	Boron (µg/L)	Ratio of boron to lithium	Reference
USGS 23	434055112595901	05-21-91	6	NS	NA	Liszewski and Mann, 1993
		04-19-95	4.3	26	6	Busenberg and others, 2000
		10-15-96	4.2	26	6	Busenberg and others, 2000
USGS 26	435212112394001	05-23-91	17	NS	NA	Liszewski and Mann, 1993
		10-15-96	18.4	38	2	Busenberg and others, 2000
USGS 27	434851112321801	10-15-96	36.4	52	1	Busenberg and others, 2000
USGS 29	434407112285101	06-12-91	26	NS	NA	Liszewski and Mann, 1993
		06-15-95	23.7	36	2	Busenberg and others, 2000
		07-19-96	24	36	2	Busenberg and others, 2000
USGS 31	434625112342101	06-12-91	18	NS	NA	Liszewski and Mann, 1993
		06-15-95	17.8	35	2	Busenberg and others, 2000
		07-19-96	18.1	37	2	Busenberg and others, 2000
USGS 32	434444112322101	06-12-91	21	NS	NA	Liszewski and Mann, 1993
		06-15-95	19.1	43	2	Busenberg and others, 2000
		07-16-96	18.1	38	2	Busenberg and others, 2000
USGS 36	433330112565201	07-16-96	1.7	21	12	Busenberg and others, 2000
USGS 82	433401112551001	07-16-96	2.2	19	9	Busenberg and others, 2000
USGS 83	433023112561501	04-11-95	3	15	5	Busenberg and others, 2000
USGS 86	432935113080001	10-11-96	2.3	18	8	Busenberg and others, 2000
USGS 89	433005113032801	07-17-96	4.2	30	7	Busenberg and others, 2000
USGS 97	433807112551501	06-13-95	2.6	29	11	Busenberg and others, 2000
USGS 98	433657112563601	06-12-95	2.5	22	9	Busenberg and others, 2000
USGS 99	433705112552101	06-12-95	2.5	30	12	Busenberg and others, 2000
USGS 100	433503112400701	04-21-95	23.4	44	2	Busenberg and others, 2000
		10-10-96	22.6	46	2	Busenberg and others, 2000
USGS 101	433255112381801	05-15-91	28	NS	NA	Liszewski and Mann, 1993
		04-21-95	27.7	45	2	Busenberg and others, 2000
		10-10-96	27.8	47	2	Busenberg and others, 2000
USGS 102	433853112551601	06-13-95	2.9	30	10	Busenberg and others, 2000

Appendix E. Concentrations of lithium and boron in water from sampling sites located at the Idaho National Laboratory and vicinity, Idaho.—Continued

[**Local name:** is the local well identifier used in this study. **Site identifier:** the unique numerical identifiers used to access well data within the National Water Information System (NWIS, http://waterdata.usgs.gov/nwis). **Lithium and boron:** concentrations in micrograms per liter. **Abbreviations:** µg/L, micrograms per liter; NS, not sampled; NA, not applicable; –, not available]

Local name	Site identifier	Date sampled	Lithium (µg/L)	Boron (µg/L)	Ratio of boron to lithium	Reference
USGS 103	432714112560701	04-18-95	6.9	29	4	Busenberg and others, 2000
		07-15-96	6.9	30	4	Busenberg and others, 2000
	432714112560723	10-02-07	5.14	NS	NA	Bartholomay and Twining, 2010
		08-20-08	6.26	NS	NA	Bartholomay and Twining, 2010
	432714112560720	10-02-07	3.11	NS	NA	Bartholomay and Twining, 2010
		08-19-08	4.32	NS	NA	Bartholomay and Twining, 2010
	432714112560716	10-01-07	2.23	NS	NA	Bartholomay and Twining, 2010
		08-19-08	3.92	NS	NA	Bartholomay and Twining, 2010
	432714112560712	10-01-07	1.58	NS	NA	Bartholomay and Twining, 2010
		08-18-08	2.32	NS	NA	Bartholomay and Twining, 2010
	432714112560708	10-01-07	1.51	NS	NA	Bartholomay and Twining, 2010
		08-18-08	2.02	NS	NA	Bartholomay and Twining, 2010
	432714112560704	09-25-07	2.45	NS	NA	Bartholomay and Twining, 2010
		08-18-08	2.03	NS	NA	Bartholomay and Twining, 2010
		08-18-08	2.09	NS	NA	Bartholomay and Twining, 2010
	432714112560702	09-25-07	1.76	NS	NA	Bartholomay and Twining, 2010
		08-19-08	2.12	NS	NA	Bartholomay and Twining, 2010
USGS 104	432856112560801	04-18-95	2.4	16	7	Busenberg and others, 2000
		07-15-96	2.2	16	7	Busenberg and others, 2000
USGS 105	432703113001801	04-18-95	2.5	22	9	Busenberg and others, 2000
	432703113001818	09-18-09	1.994	NS	NA	U.S. Geological Survey, 2011
	432703113001815	09-18-09	2.091	NS	NA	U.S. Geological Survey, 2011
	432703113001811	09-17-09	2.06	NS	NA	U.S. Geological Survey, 2011
	432703113001807	09-17-09	2.05	NS	NA	U.S. Geological Survey, 2011
	432703113001803	09-16-09	2.037	NS	NA	U.S. Geological Survey, 2011
USGS 107	432942112532801	10-09-96	10.5	35	3	Busenberg and others, 2000
USGS 108	432659112582601	04-19-95	4.3	23	5	Busenberg and others, 2000
	432659112582616	09-21-10	2.913	NS	NA	U.S. Geological Survey, 2011
		06-23-11	3.741	NS	NA	U.S. Geological Survey, 2011
	432659112582613	09-21-10	2.341	NS	NA	U.S. Geological Survey, 2011
		06-23-11	2.329	NS	NA	U.S. Geological Survey, 2011
	432659112582610	09-20-10	2.265	NS	NA	U.S. Geological Survey, 2011
		06-23-11	2.185	NS	NA	U.S. Geological Survey, 2011
	432659112582606	09-22-10	2.371	NS	NA	U.S. Geological Survey, 2011
		06-22-11	2.297	NS	NA	U.S. Geological Survey, 2011
	432659112582602	09-20-10	2.613	NS	NA	U.S. Geological Survey, 2011
		06-22-11	2.638	NS	NA	U.S. Geological Survey, 2011
USGS 109	432701113025601	04-20-95	3	20	7	Busenberg and others, 2000
		10-11-96	2.7	23	9	Busenberg and others, 2000
USGS 110A	432717112501502	05-08-91	17	NS	NA	Liszewski and Mann, 1993
		10-09-96	15.9	38	2	Busenberg and others, 2000

Appendix E.   Concentrations of lithium and boron in water from sampling sites located at the Idaho National Laboratory and vicinity, Idaho.—Continued

[**Local name:** is the local well identifier used in this study. **Site identifier:** the unique numerical identifiers used to access well data within the National Water Information System (NWIS, http://waterdata.usgs.gov/nwis). **Lithium and boron:** concentrations in micrograms per liter. **Abbreviations:** µg/L, micrograms per liter; NS, not sampled; NA, not applicable; –, not available]

Local name	Site identifier	Date sampled	Lithium (µg/L)	Boron (µg/L)	Ratio of boron to lithium	Reference
USGS 112	433314112563001	10-15-91	5	NS	NA	Liszewski and Mann, 1993
		07-15-96	2.4	24	10	Busenberg and others, 2000
USGS 113	433314112561801	07-16-96	2.8	26	9	Busenberg and others, 2000
USGS 115	433320112554101	07-15-96	2.1	18	9	Busenberg and others, 2000
USGS 116	433331112553201	07-15-96	2.5	18	7	Busenberg and others, 2000
USGS 117	432955113025901	10-16-91	7	NS	NA	Liszewski and Mann, 1993
		07-17-96	5.1	23	5	Busenberg and others, 2000
USGS 120	432919113031501	10-25-91	6	NS	NA	Liszewski and Mann, 1993
		07-17-96	3.6	39	11	Busenberg and others, 2000
USGS 124	432307112583101	04-20-95	6.9	20	3	Busenberg and others, 2000
		10-09-96	6.7	21	3	Busenberg and others, 2000
USGS 125	432602113052801	06-16-95	3.2	21	7	Busenberg and others, 2000
		10-11-96	3.1	22	7	Busenberg and others, 2000
Wagoner Ranch	440813112532201	05-22-97	3.3	18	5	Busenberg and others, 2000
BLR Mackay Dam	13127000	06-28-95	1.7	12	7	Busenberg and others, 2000
BLR Mackay Brdg	13127780	06-17-95	1.7	11	6	Busenberg and others, 2000
BLR Lincoln Blvd	13132535	06-19-95	2	13	7	Busenberg and others, 2000
Birch Crk at Blue Dome	13117020	06-19-95	3.2	14	4	Busenberg and others, 2000
		06-17-95	3.1	14	5	Busenberg and others, 2000
Camas Crk Mud Lake	13115000	06-17-95	3.4	14	4	Busenberg and others, 2000
Lidy Hot Springs	440832112331001	07-20-96	42.8	88	2	Busenberg and others, 2000
		05-14-97	43.3	81	2	Busenberg and others, 2000
LLR near INEEL	–	06-17-95	2.4	37	15	Busenberg and others, 2000
LLR north of Howe	13119000	06-28-95	1.1	12	11	Busenberg and others, 2000
Stoddart	435402112332101	06-12-91	47	NS	NA	Liszewski and Mann, 1993
		–	47	NS	NA	Knobel and others, 1999
Reno Ranch	440142112425501	–	25	NS	NA	Swanson and others, 2003
USGS 126 A	435529112471301	–	4.7	NS	NA	Swanson and others, 2003
USGS 126 B	435529112471401	–	5.5	NS	NA	Swanson and others, 2003
6N-34E-32ACD1	434818112284801	08-29-89	32	50	2	Spinazola and others, 1992
6N-35E-12BCD1	435150112172701	08-29-89	18	60	3	Spinazola and others, 1992
6N-35E-21AAB1	435028112202601	08-29-89	24	30	1	Spinazola and others, 1992
7N-33E-16BAB1	435632112351701	08-29-89	41	120	3	Spinazola and others, 1992
7N-34E-10ACA1	435712112263201	08-29-89	14	30	2	Spinazola and others, 1992
7N-35E-22DAD1	435505112190201	08-30-89	14	20	1	Spinazola and others, 1992
7N-36E-5CAA1	435753112145101	08-30-89	16	20	1	Spinazola and others, 1992
8N-37E-27BBC1	435949112054501	08-30-89	26	60	2	Spinazola and others, 1992

Appendix E.   Concentrations of lithium and boron in water from sampling sites located at the Idaho National Laboratory and vicinity, Idaho.—Continued

[**Local name:** is the local well identifier used in this study. **Site identifier:** the unique numerical identifiers used to access well data within the National Water Information System (NWIS, http://waterdata.usgs.gov/nwis). **Lithium and boron:** concentrations in micrograms per liter. **Abbreviations:** μg/L, micrograms per liter; NS, not sampled; NA, not applicable; –, not available]

Local name	Site identifier	Date sampled	Lithium (μg/L)	Boron (μg/L)	Ratio of boron to lithium	Reference
8N-37E-30ABC1	435951112084701	08-30-89	21	50	2	Spinazola and others, 1992
Grazing Well 2	431553112492001	06-21-93	18	NS	NA	Bartholomay and others, 2001
Grazing Service CCC 3	430911112585401	06-21-93	17	NS	NA	Bartholomay and others, 2001
Houghland Well	431439113071401	06-22-93	12	NS	NA	Bartholomay and others, 2001
Crossroads Well	432128113092701	06-22-93	4	NS	NA	Bartholomay and others, 2001
Fingers Butte Well	432424113165301	06-22-93	4	NS	NA	Bartholomay and others, 2001
USGS 20	433253112545901	05-30-91	6	NS	NA	Liszewski and Mann, 1993
		09-06-77	NS	20	NA	U.S. Geological Survey, 2011
USGS 57	433344112562601	05-13-91	4	NS	NA	Liszewski and Mann, 1993
USGS 65	433447112574501	05-16-91	5	NS	NA	Liszewski and Mann, 1993
USGS 85	433246112571201	06-04-91	6	NS	NA	Liszewski and Mann, 1993
USGS 88	432940113030201	10-15-91	9	NS	NA	Liszewski and Mann, 1993
USGS 119	432945113023401	10-15-91	5	NS	NA	Liszewski and Mann, 1993
USGS 121	433450112560301	10-15-91	5	NS	NA	Liszewski and Mann, 1993
USGS 122	433353112555201	10-15-91	5	NS	NA	Liszewski and Mann, 1993
		10-15-91	6	NS	NA	Liszewski and Mann, 1993
USGS 123	433352112561401	10-15-91	6	NS	NA	Liszewski and Mann, 1993
CPP Pond 1	433351112555101	06-06-91	11	NS	NA	Liszewski and Mann, 1993
McKinney (10N 29E 24AAD1)	441113112560601	06-13-91	5	NS	NA	Liszewski and Mann, 1993
No Name 1	435038112453401	05-22-91	5	NS	NA	Liszewski and Mann, 1993
Simplot 1 (05N 29E 01BBB1)	434751112571801	06-11-91	5	NS	NA	Liszewski and Mann, 1993
ANP 10	434909112400401	09-07-77	NS	30	NA	U.S. Geological Survey, 2011
FET 1	435120112432101	04-17-58	NS	50	NA	U.S. Geological Survey, 2011
FET 2	435119112431801	05-03-58	NS	40	NA	U.S. Geological Survey, 2011
OMRE	433116112534701	03-23-57	NS	110	NA	U.S. Geological Survey, 2011
SPERT 1	433252112520301	02-27-56	NS	50	NA	U.S. Geological Survey, 2011
		09-06-77	NS	20	NA	U.S. Geological Survey, 2011
USGS 30	434601112315401	04-22-53	NS	80	NA	U.S. Geological Survey, 2011
		04-27-53	NS	60	NA	U.S. Geological Survey, 2011

Appendix E.   Concentrations of lithium and boron in water from sampling sites located at the Idaho National Laboratory and vicinity, Idaho.—Continued

[**Local name:** is the local well identifier used in this study. **Site identifier:** the unique numerical identifiers used to access well data within the National Water Information System (NWIS, http://waterdata.usgs.gov/nwis). **Lithium and boron:** concentrations in micrograms per liter. **Abbreviations:** µg/L, micrograms per liter; NS, not sampled; NA, not applicable; –, not available]

Local name	Site identifier	Date sampled	Lithium (µg/L)	Boron (µg/L)	Ratio of boron to lithium	Reference
USGS 132	432906113025022	09-06-06	2.49	NS	NA	Bartholomay and Twining, 2010
		09-18-07	2.54	NS	NA	Bartholomay and Twining, 2010
		08-14-08	2.86	NS	NA	Bartholomay and Twining, 2010
		08-14-08	2.90	NS	NA	Bartholomay and Twining, 2010
	432906113025018	09-05-06	2.16	NS	NA	Bartholomay and Twining, 2010
		09-18-07	2.12	NS	NA	Bartholomay and Twining, 2010
		08-13-08	2.16	NS	NA	Bartholomay and Twining, 2010
	432906113025014	09-05-06	1.95	NS	NA	Bartholomay and Twining, 2010
		09-18-07	2.28	NS	NA	Bartholomay and Twining, 2010
		08-13-08	1.78	NS	NA	Bartholomay and Twining, 2010
	432906113025010	08-31-06	2.17	NS	NA	Bartholomay and Twining, 2010
		09-18-07	2.28	NS	NA	Bartholomay and Twining, 2010
		08-13-08	1.86	NS	NA	Bartholomay and Twining, 2010
	432906113025006	08-30-06	2.05	NS	NA	Bartholomay and Twining, 2010
		09-17-07	2.17	NS	NA	Bartholomay and Twining, 2010
		08-12-08	1.73	NS	NA	Bartholomay and Twining, 2010
	432906113025001	08-29-06	2.05	NS	NA	Bartholomay and Twining, 2010
		09-17-07	2.32	NS	NA	Bartholomay and Twining, 2010
		08-12-08	1.78	NS	NA	Bartholomay and Twining, 2010
USGS 133	433605112554312	09-24-07	1.8	NS	NA	Bartholomay and Twining, 2010
		09-02-2008	0.66	NS	NA	Bartholomay and Twining, 2010
	433605112554308	09-24-07	1.85	NS	NA	Bartholomay and Twining, 2010
		09-02-08	0.85	NS	NA	Bartholomay and Twining, 2010
	433605112554305	09-24-07	2.28	NS	NA	Bartholomay and Twining, 2010
		09-09-08	1.74	NS	NA	Bartholomay and Twining, 2010
	433605112554301	09-24-07	2.58	NS	NA	Bartholomay and Twining, 2010
		09-02-08	1.37	NS	NA	Bartholomay and Twining, 2010
USGS 134	433611112595819	09-27-06	2.95	NS	NA	Bartholomay and Twining, 2010
		09-10-07	2.56	NS	NA	Bartholomay and Twining, 2010
		09-04-08	2.03	NS	NA	Bartholomay and Twining, 2010
	433611112595815	09-28-06	3.33	NS	NA	Bartholomay and Twining, 2010
		09-06-07	2.96	NS	NA	Bartholomay and Twining, 2010
		09-06-07	2.98	NS	NA	Bartholomay and Twining, 2010
		09-04-08	2.9	NS	NA	Bartholomay and Twining, 2010
	433611112595811	09-27-06	1.56	NS	NA	Bartholomay and Twining, 2010
		09-05-07	2.15	NS	NA	Bartholomay and Twining, 2010
		09-03-08	2.32	NS	NA	Bartholomay and Twining, 2010
	433611112595807	09-26-06	2.5	NS	NA	Bartholomay and Twining, 2010
		09-05-07	2.35	NS	NA	Bartholomay and Twining, 2010
		09-03-08	1.86	NS	NA	Bartholomay and Twining, 2010
	433611112595804	09-03-08	2.71	NS	NA	Bartholomay and Twining, 2010
	433611112595803	09-25-06	3.58	NS	NA	Bartholomay and Twining, 2010
		09-04-07	3.54	NS	NA	Bartholomay and Twining, 2010

Appendix E.   Concentrations of lithium and boron in water from sampling sites located at the Idaho National Laboratory and vicinity, Idaho.—Continued

[**Local name:** is the local well identifier used in this study. **Site identifier:** the unique numerical identifiers used to access well data within the National Water Information System (NWIS, http://waterdata.usgs.gov/nwis). **Lithium and boron:** concentrations in micrograms per liter. **Abbreviations:** µg/L, micrograms per liter; NS, not sampled; NA, not applicable; –, not available]

Local name	Site identifier	Date sampled	Lithium (µg/L)	Boron (µg/L)	Ratio of boron to lithium	Reference
USGS 135	432753113093613	09-15-09	1.633	NS	NA	U.S. Geological Survey, 2011
	432753113093609	09-15-09	1.589	NS	NA	U.S. Geological Survey, 2011
	432753113093605	09-15-09	1.431	NS	NA	U.S. Geological Survey, 2011
	432753113093601	09-14-09	2.052	NS	NA	U.S. Geological Survey, 2011
MIDDLE 2050A	433409112570515	09-19-06	1.62	NS	NA	Bartholomay and Twining, 2010
		09-20-07	1.32	NS	NA	Bartholomay and Twining, 2010
		08-27-08	1.37	NS	NA	Bartholomay and Twining, 2010
		08-27-08	1.34	NS	NA	Bartholomay and Twining, 2010
	433409112570512	09-19-06	2.04	NS	NA	Bartholomay and Twining, 2010
		09-20-07	1.35	NS	NA	Bartholomay and Twining, 2010
		08-27-08	1.47	NS	NA	Bartholomay and Twining, 2010
	433409112570509	09-20-06	2.45	NS	NA	Bartholomay and Twining, 2010
		09-20-06	2.46	NS	NA	Bartholomay and Twining, 2010
		09-20-07	1.90	NS	NA	Bartholomay and Twining, 2010
		08-26-08	1.98	NS	NA	Bartholomay and Twining, 2010
	433409112570506	09-18-06	3.08	NS	NA	Bartholomay and Twining, 2010
		09-19-07	3.00	NS	NA	Bartholomay and Twining, 2010
		08-26-08	2.19	NS	NA	Bartholomay and Twining, 2010
	433409112570503	09-18-06	4.15	NS	NA	Bartholomay and Twining, 2010
		09-19-07	3.51	NS	NA	Bartholomay and Twining, 2010
		08-26-08	3.25	NS	NA	Bartholomay and Twining, 2010
MIDDLE 2051	433217113004912	09-11-06	1.23	NS	NA	Bartholomay and Twining, 2010
		09-12-07	1.12	NS	NA	Bartholomay and Twining, 2010
		08-25-08	1.59	NS	NA	Bartholomay and Twining, 2010
	433217113004909	09-13-06	2.16	NS	NA	Bartholomay and Twining, 2010
		09-13-06	2.17	NS	NA	Bartholomay and Twining, 2010
		09-12-07	1.53	NS	NA	Bartholomay and Twining, 2010
		09-12-07	1.47	NS	NA	Bartholomay and Twining, 2010
		08-25-08	1.88	NS	NA	Bartholomay and Twining, 2010
	433217113004906	09-12-06	2.14	NS	NA	Bartholomay and Twining, 2010
		09-12-06	2.06	NS	NA	Bartholomay and Twining, 2010
		09-11-07	1.84	NS	NA	Bartholomay and Twining, 2010
		08-21-08	1.90	NS	NA	Bartholomay and Twining, 2010
	433217113004903	09-11-06	2.74	NS	NA	Bartholomay and Twining, 2010
		09-11-07	2.49	NS	NA	Bartholomay and Twining, 2010
		08-21-08	2.06	NS	NA	Bartholomay and Twining, 2010
	433217113004901	09-07-06	2.84	NS	NA	Bartholomay and Twining, 2010
		09-11-07	2.60	NS	NA	Bartholomay and Twining, 2010
		08-21-08	2.18	NS	NA	Bartholomay and Twining, 2010

Appendix F. Binary Mixing Equation in Ratio-Element Space

End Members

Binary mixing models commonly are used to explain variations in geochemical data. The mixing process combines distinct sources (end members) to make a range of products with intermediate compositions. The subregional groundwater flow model of the eastern Snake River Plain aquifer has two major water sources: tributary valley (*tv*) underflow across the northwest mountain-front boundary and regional aquifer (*ra*) underflow across the northeast boundary. The mixture of these two sources is referred to as *mix*. The measured composition of water types within the model domain are summarized in table F1. The trace elements boron and lithium are considered in the mixing equation. Large standard deviations in the water composition of both end members indicate varying degrees of mixture throughout the model domain; these chemical heterogeneities were accounted for by using average values for each end member in the mixing model (table F1).

Table F1. Composition of groundwater in the tributary valley and regional aquifer, measured from wells at the Idaho National Laboratory and vicinity, Idaho.

[Measurements are reported as the arithmetic mean plus or minus one standard deviation. Boron and lithium concentrations are in micrograms per liter]

Composition	Tributary valley water (*tv*) (n = 48)	Regional aquifer water (*ra*) (n = 39)
Boron concentration, [B]	22 ± 6	43 ± 19
Lithium concentration, [Li]	2.9 ± 0.9	20.8 ± 12.3
Boron to lithium ratio, (B/Li)	8 ± 2	2 ± 1

Element-Element Space

For two elements, boron and lithium, conservation of atoms and mass is in the form

$$[B]_{mix} = f_{ra}[B]_{ra} + f_{tv}[B]_{tv}$$
$$[Li]_{mix} = f_{ra}[Li]_{ra} + f_{tv}[Li]_{tv}$$

(1)

where f_{tv} and f_{ra} are the mass fraction of tributary-valley water and regional-aquifer water, respectively, such that $f_{tv} + f_{ra} = 1$. Equation (1) is then rewritten as:

$$[B]_{mix} = f_{ra}[B]_{ra} + (1 - f_{ra})[B]_{tv}$$
$$[Li]_{mix} = f_{ra}[Li]_{ra} + (1 - f_{ra})[Li]_{tv}$$

(2)

A large variability in boron concentrations indicates that the boron element is ill suited for the mixing model; therefore, the more stable boron to lithium ratio is used in its place.

Element-Ratio Space

The boron to lithium ratio of mixtures is obtained by taking the ratio of the conservation equations (eq. 2),

$$(B/Li)_{mix} = \frac{f_{ra}[B]_{ra} + (1-f_{ra})[B]_{tv}}{f_{ra}[Li]_{ra} + (1-f_{ra})[Li]_{tv}} \tag{3}$$

and substituting [B] = (B/LI) [Li] for *tv* and *ra*:

$$(B/Li)_{mix} = \frac{f_{ra}[Li]_{ra}(B/Li)_{ra} + (1-f_{ra})[Li]_{tv}(B/Li)_{tv}}{f_{ra}[Li]_{ra} + (1-f_{ra})[Li]_{tv}} \tag{4}$$

where the mixture ratio is weighted by the concentration of lithium in each end member. From equation (4), the mass fraction of regional aquifer water in the mixing zone can be expressed as:

$$(B/Li)_{mix} = \frac{f_{ra}[Li]_{ra}(B/Li)_{ra} + (1-f_{ra})[Li]_{tv}(B/Li)_{tv}}{f_{ra}[Li]_{ra} + (1-f_{ra})[Li]_{tv}}$$

$$f_{ra}[Li]_{ra}(B/Li)_{mix} + [Li]_{tv}(B/Li)_{mix} - f_{ra}[Li]_{tv}(B/Li)_{mix} =$$
$$f_{ra}[Li]_{ra}(B/Li)_{ra} + [Li]_{tv}(B/Li)_{tv} - f_{ra}[Li]_{tv}(B/Li)_{tv}$$

$$f_{ra}[Li]_{ra}(B/Li)_{mix} - f_{ra}[Li]_{tv}(B/Li)_{mix} - f_{ra}[Li]_{ra}(B/Li)_{ra} + f_{ra}[Li]_{tv}(B/Li)_{tv} =$$
$$[Li]_{tv}(B/Li)_{tv} - [Li]_{tv}(B/Li)_{mix}$$

$$f_{ra}\left([Li]_{ra}(B/Li)_{mix} - [Li]_{tv}(B/Li)_{mix} - [Li]_{ra}(B/Li)_{ra} + [Li]_{tv}(B/Li)_{tv}\right) =$$
$$[Li]_{tv}\left((B/Li)_{tv} - (B/Li)_{mix}\right)$$

$$f_{ra} = \frac{[Li]_{tv}\left((B/Li)_{tv} - (B/Li)_{mix}\right)}{[Li]_{ra}\left((B/Li)_{mix} - (B/Li)_{ra}\right) + [Li]_{tv}\left((B/Li)_{tv} - (B/Li)_{mix}\right)} \tag{5}$$

Validity of Mixing Model

A functional relation between the boron to lithium ratio and an arbitrary elemental concentration, [X], is obtained by replacing the fraction of regional aquifer water in equation (4) by

$$[X]_{mix} = f_{ra}[X]_{ra} + (1-f_{ra})[X]_{tv} \tag{6}$$
$$f_{ra} = \frac{[X]_{mix} - [X]_{tv}}{[X]_{ra} - [X]_{tv}}$$

Equation (4) is then rewritten as:

$$(B/Li)_{mix} = \frac{f_{ra}[Li]_{ra}(B/Li)_{ra} + (1-f_{ra})[Li]_{tv}(B/Li)_{tv}}{f_{ra}[Li]_{ra} + (1-f_{ra})[Li]_{tv}}$$

$$(B/Li)_{mix} = \frac{\left(\dfrac{[X]_{mix}-[X]_{tv}}{[X]_{ra}-[X]_{tv}}\right)[Li]_{ra}(B/Li)_{ra} + \left(1-\left(\dfrac{[X]_{mix}-[X]_{tv}}{[X]_{ra}-[X]_{tv}}\right)\right)[Li]_{tv}(B/Li)_{tv}}{\left(\dfrac{[X]_{mix}-[X]_{tv}}{[X]_{ra}-[X]_{tv}}\right)[Li]_{ra} + \left(1-\left(\dfrac{[X]_{mix}-[X]_{tv}}{[X]_{ra}-[X]_{tv}}\right)\right)[Li]_{tv}}$$

$$(B/Li)_{mix} = \frac{\dfrac{[X]_{mix}[Li]_{ra}(B/Li)_{ra}}{[X]_{ra}-[X]_{tv}} - \dfrac{[X]_{tv}[Li]_{ra}(B/Li)_{ra}}{[X]_{ra}-[X]_{tv}} +}{\dfrac{[X]_{mix}[Li]_{ra}}{[X]_{ra}-[X]_{tv}} - \dfrac{[X]_{tv}[Li]_{ra}}{[X]_{ra}-[X]_{tv}} -}$$

$$\frac{\dfrac{[X]_{ra}[Li]_{tv}(B/Li)_{tv}}{[X]_{ra}-[X]_{tv}} - \dfrac{[X]_{tv}[Li]_{tv}(B/Li)_{tv}}{[X]_{ra}-[X]_{tv}} - \dfrac{[X]_{mix}[Li]_{tv}(B/Li)_{tv}}{[X]_{ra}-[X]_{tv}} + \dfrac{[X]_{tv}[Li]_{tv}(B/Li)_{tv}}{[X]_{ra}-[X]_{tv}}}{\dfrac{[X]_{ra}[Li]_{tv}}{[X]_{ra}-[X]_{tv}} - \dfrac{[X]_{tv}[Li]_{tv}}{[X]_{ra}-[X]_{tv}} - \dfrac{[X]_{mix}[Li]_{tv}}{[X]_{ra}-[X]_{tv}} + \dfrac{[X]_{tv}[Li]_{tv}}{[X]_{ra}-[X]_{tv}}}$$

$$(B/Li)_{mix} =$$

$$\frac{[X]_{mix}[Li]_{ra}(B/Li)_{ra} - [X]_{tv}[Li]_{ra}(B/Li)_{ra} + [X]_{ra}[Li]_{tv}(B/Li)_{tv} - [X]_{mix}[Li]_{tv}(B/Li)_{tv}}{[X]_{mix}[Li]_{ra} - [X]_{tv}[Li]_{ra} + [X]_{ra}[Li]_{tv} - [X]_{mix}[Li]_{tv}}$$

$$[X]_{mix}[Li]_{ra}(B/Li)_{mix} - [X]_{tv}[Li]_{ra}(B/Li)_{mix} + [X]_{ra}[Li]_{tv}(B/Li)_{mix} -$$
$$[X]_{mix}[Li]_{tv}(B/Li)_{mix} = [X]_{mix}[Li]_{ra}(B/Li)_{ra} - [X]_{tv}[Li]_{ra}(B/Li)_{ra} +$$
$$[X]_{ra}[Li]_{tv}(B/Li)_{tv} - [X]_{mix}[Li]_{tv}(B/Li)_{tv}$$

$$\left([X]_{ra}[Li]_{tv} - [X]_{tv}[Li]_{ra}\right)(B/Li)_{mix} + \left([Li]_{ra} - [Li]_{tv}\right)[X]_{mix}(B/Li)_{mix} +$$
$$\left([Li]_{tv}(B/Li)_{tv} - [Li]_{ra}(B/Li)_{ra}\right)[X]_{mix} + \left([X]_{tv}[Li]_{ra}(B/Li)_{ra} - [X]_{ra}[Li]_{tv}(B/Li)_{tv}\right) = 0$$

$$A(B/Li)_{mix} + B[X]_{mix}(B/Li)_{mix} + C[X]_{mix} + D = 0 \qquad (7)$$

where

$$A = [X]_{ra}[Li]_{tv} - [X]_{tv}[Li]_{ra} \qquad (8)$$
$$B = [Li]_{ra} - [Li]_{tv}$$
$$C = [Li]_{tv}(B/Li)_{tv} - [Li]_{ra}(B/Li)_{ra}$$
$$D = [X]_{tv}[Li]_{ra}(B/Li)_{ra} - [X]_{ra}[Li]_{tv}(B/Li)_{tv}$$

Substituting [Li] for [X] in equations (7) and (8) gives the mixing curve in element-ratio space as a hyperbola (Langmuir and others, 1978, p. 381, equation 1),

$$A(B/Li)_{mix} + B[Li]_{mix}(B/Li)_{mix} + C[Li]_{mix} + D = 0 \qquad (9)$$

where

$$
\begin{aligned}
A &= [Li]_{ra}[Li]_{tv} - [Li]_{tv}[Li]_{ra} \\
B &= [Li]_{ra} - [Li]_{tv} \\
C &= [Li]_{tv}(B/Li)_{tv} - [Li]_{ra}(B/Li)_{ra} \\
D &= [Li]_{tv}[Li]_{ra}(B/Li)_{ra} - [Li]_{ra}[Li]_{tv}(B/Li)_{tv}
\end{aligned}
\qquad (10)
$$

The mixing model based on average end-member concentrations and concentration ratios is calculated by substituting average values (table F1) into equation (10),

$$
\begin{aligned}
A &= 20.8\,\mu g/L \times 2.9\,\mu g/L - 2.9\,\mu g/L \times 20.8\,\mu g/L = 0 \\
B &= 20.8\,\mu g/L - 2.9\,\mu g/L = 17.9 \\
C &= 2.9\,\mu g/L \times 8 - 20.8\,\mu g/L \times 2 = -18 \\
D &= 2.9\,\mu g/L \times 20.8\,\mu g/L \times 2 - 20.8\,\mu g/L \times 2.9\,\mu g/L \times 8 = -362
\end{aligned}
\qquad (11)
$$

Equation (9) is then expressed as

$$17.9[Li]_{mix}(B/Li)_{mix} - 18[Li]_{mix} - 362 = 0 \qquad (12)$$

The mixing model does a moderate job of fitting the observational data with a 0.71 coefficient of determination (R^2). The observational data and mixing curve are shown in element-ratio space in figure F1.

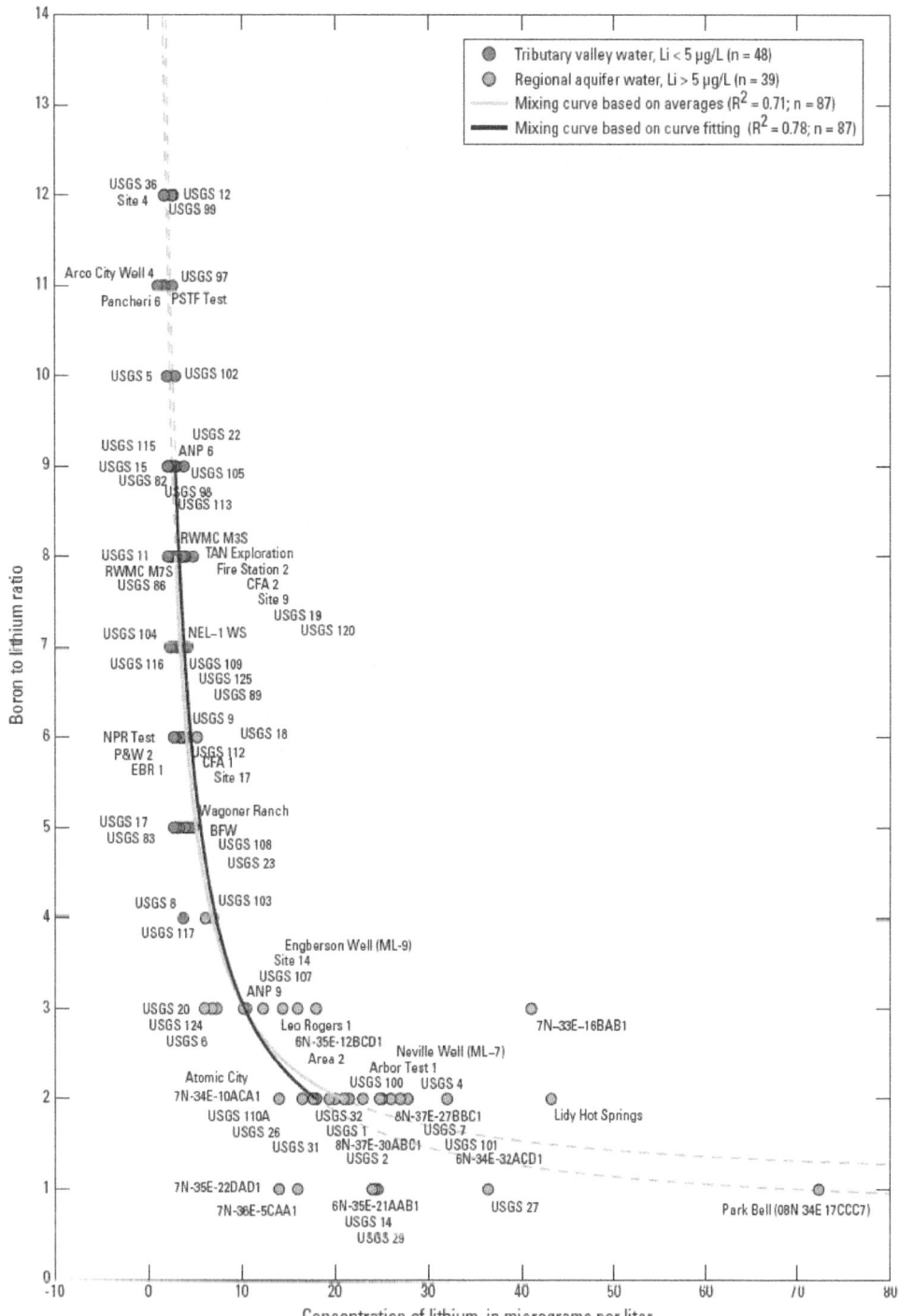

Figure F1. Data and mixing curve in element-ratio space.

End member compositions are not necessary for calculating the mixing curve. Alternatively, the mixing curve may be determined using curve fitting, a minimization of the root-mean-square error (RMSE) by systematically choosing real numbers (**R**) for end-members in the mixing model (eq. 9). The unconstrained optimization problem is

$$\min_{[\text{Li}_{ra}],(\text{B/Li}_{tv})\in\mathbf{R}}\left(\sqrt{\frac{\sum_i\left([\mathbf{Li}]_{\text{obs},i}-[\mathbf{Li}]_{\text{calc},i}\right)^2}{n_{\text{Li}<5}}}+\sqrt{\frac{\sum_j\left((\mathbf{B/Li})_{\text{obs},j}-(\mathbf{B/Li})_{\text{calc},j}\right)^2}{n_{\text{Li}>5}}}\right)\tag{13}$$

subject to:

$$[\text{Li}]_{\text{calc},i}=\frac{-A(\text{B/Li})_{\text{calc},i}-D}{B+C(\text{B/Li})_{\text{calc},i}}\tag{14}$$

$$(\text{B/Li})_{\text{calc},j}=\frac{-C[\text{Li}]_{\text{obs},j}-D}{A+B[\text{Li}]_{\text{obs},j}}$$

where $n_{\text{Li}<5}$ is the number of data records with Li values less than 5 µg/L (n = 48), $n_{\text{Li}>5}$ is the number of data records with Li values greater than 5 µg/L (n = 39), i is an index number for data records with Li values less than 5 µg/L, j is an index number for data records with Li values greater than 5 µg/L, $[\text{Li}]_{\text{obs}}$ is an observed lithium concentration, $[\text{Li}]_{\text{calc}}$ is the calculated lithium concentration, $(\text{B/Li})_{\text{obs}}$ is a measured boron to lithium concentration ratio, $(\text{B/Li})_{\text{calc}}$ is the calculated boron to lithium concentration ratio, and A, B, C, and D are the coefficients of the mixing model defined in equation (10). Limiting the number of decision variables to two allowed for a unique solution to the optimization problem (eq. 13). End members with the highest standard deviation, $[\text{Li}]_{ra}$ ($\sigma=12.3$ µg/L) and $(\text{B/Li})_{tv}$ ($\sigma=2$), were selected as the decision variables, and end-members with the lowest standard deviations, $[\text{Li}]_{tv}$ ($\sigma=0.9$ µg/L) and $(\text{B/Li})_{ra}$ ($\sigma=1$), were set constant at their average values. The resulting mixing curve based on curve fitting ($R^2 = 0.78$) is shown in figure F1 and expressed as

$$10.9[\text{Li}]_{mix}(\text{B/Li})_{mix}-11[\text{Li}]_{mix}-192=0\tag{15}$$

The end member compositions based on the solution to equation (13) (RMSE = 1.47) are

$$[Li]_{tv}=2.9\,\text{µg/L}\tag{16}$$
$$[Li]_{ra}^*=14\,\text{µg/L}$$
$$(\text{B/Li})_{tv}^*=7$$
$$(B/Li)_{ra}=2$$

where superscript * indicates an optimal value for the decision variable. At lithium concentrations of 2.9 and 14 µg/L, the mass fraction of regional water within the mixing zone is 0 and 1, respectively.

Sensitivity of the optimal solution to changes in the fixed end members was analyzed varying $[Li]_{tv}$ and $(B/Li)_{ra}$ one standard deviation from its mean (table F2). Results of the sensitivity analysis indicate that $[Li]^*_{ra}$ is insensitive to changes in $[Li]_{tv}$ and highly sensitive to changes in $(B/Li)_{ra}$; whereas, $(B/Li)_{tv}$ is moderately sensitive to changes in $[Li]_{tv}$ and $(B/Li)_{ra}$.

Table F2. Sensitivity of the optimal solution to changes in the end member composition of lithium in the tributary valley water and the boron to lithium concentration ratio in regional aquifer water.

[**Abbreviations:** $[Li]_{tv}$, end-member lithium concentration of tributary valley water in micrograms per liter; $[Li]^*_{ra}$, optimal end-member lithium concentration of regional aquifer water in micrograms per liter; $(B/Li)^*_{tv}$, optimal end-member boron to lithium concentration ratio of tributary valley water; $(B/Li)_{ra}$, end-member boron to lithium concentration ratio of regional aquifer water; –, infeasible solution to the optimization problem]

$[Li]_{tv}$	$[Li]^*_{ra}$	$(B/Li)^*_{tv}$	$(B/Li)_{ra}$
[3] 2.1	18	9	[1] 2
[1] 2.9	18	7	[1] 2
[2] 3.8	18	6	[1] 2
[2] 3.8	9	6	[2] 3
[3] 2.1	–	–	[3] 1
[1] 2.9	–	–	[3] 1
[1] 2.9	9	7	[2] 3

[1] Arithmetic mean.

[2] Arithmetic mean plus one standard deviation.

[3] Arithmetic mean minus one standard deviation.

Reference Cited

Langmuir, C.H., Vocke, R.D., Jr., and Hanson, G.N., 1978, A general mixing equation with applications to Icelandic basalts: Earth and Planetary Science Letters, v. 37, p. 380-392.

Appendix G. Summary of Backward Particle-Tracking Results from a Sensitivity Analysis of the Source Area Contributions to Changes in Input Parameters for the Layered-Grid and Well NPR-W01 Simulations

[Locations of source areas are shown in figure 3. Orphans are identified as particles that did not terminate in one of the specified source areas. None of the source areas include model layer 6. Streamflow reaches are located between the model boundary and streamflow-gaging station (map No. 504 (600–601), within the Big Lost River spreading area (602–605), between streamflow-gaging stations 504 and 506 (606–607), within the Big Lost River sinks and playas (608–610), within the Little Lost River downstream of the model boundary (611), and within Birch Creek downstream of the model boundary (612). **FRAC:** the fraction of the total outflow from a cell contributed by an internal source; used to identify weak-source cells. **Layer:** the model layer particles are located in at their release. **Depth interval:** the depths of released particles are distributed randomly in this depth interval, in feet below the top of model layer 1. **Radius:** the aerial radius of the cylindrical volume defining particle release locations, in feet. **Abbreviation:** NA, not applicable]

Subregional layered-grid simulation; FRAC = 0.5 and Layer = 1–6 (base case)

Simulated groundwater source area	Number of particles from source area within model layer						Percentage of source area contributions within model layer					
	1	2	3	4	5	1–6	1	2	3	4	5	1–6
Northwest mountain-front boundary												
Big Lost River valley (BLR)	5,303	NA	NA	NA	NA	5,303	32.2	NA	NA	NA	NA	12.6
Little Lost River valley (LLR)	2,512	1,749	1,626	NA	NA	5,887	15.2	28.5	30.3	NA	NA	14.0
Birch Creek valley (BC)	650	684	470	853	NA	2,657	3.9	11.2	8.8	10.5	NA	6.3
Northwest mountain-front subtotal	8,465	2,433	2,096	853	NA	13,847	51.4	39.7	39.1	10.5	NA	32.9
Northeast regional-underflow boundary												
Reno Ranch section (Re)	297	274	430	554	656	2,211	1.8	4.5	8.0	6.8	11.0	5.3
Monteview section (Mo)	652	339	352	813	NA	2,156	4.0	5.5	6.5	10.0	NA	5.1
Mud Lake section (ML)	1,999	1,388	1,137	3,075	NA	7,599	12.1	22.6	21.2	37.8	NA	18.1
Terreton section (Te)	2,046	1,698	1,344	2,833	5,296	13,217	12.4	27.7	25.1	34.9	89.0	31.4
Northeast regional-underflow subtotal	4,994	3,699	3,263	7,275	5,952	25,183	30.3	60.3	60.9	89.5	100.0	59.9
Water table boundary												
Streamflow infiltration												
Big Lost River infiltration												
Stream reach 600–601	47	NA	NA	NA	NA	47	0.3	NA	NA	NA	NA	0.1
Stream reach 602–605	0	NA	NA	NA	NA	0	0.0	NA	NA	NA	NA	0.0
Stream reach 606–607	317	NA	NA	NA	NA	317	1.9	NA	NA	NA	NA	0.8
Stream reach 608–610	1,194	NA	NA	NA	NA	1,194	7.2	NA	NA	NA	NA	2.8
Little Lost River infiltration												
Stream reach 611	0	NA	NA	NA	NA	0	0.0	NA	NA	NA	NA	0.0
Birch Creek infiltration												
Stream reach 612	923	NA	NA	NA	NA	923	5.6	NA	NA	NA	NA	2.2
Streamflow-infiltration subtotal	2,481	NA	NA	NA	NA	2,481	15.1	NA	NA	NA	NA	5.9
Orphans	542	0	1	0	1	544	3.3	0.0	0.0	0.0	0.0	1.3
Total	16,482	6,132	5,360	8,128	5,953	42,055	100.0	100.0	100.0	100.0	100.0	100.0

Appendix G. Summary of backward particle-tracking results from a sensitivity analysis of the source area contributions to changes in input parameters for the layered-grid and NPR-W01 well simulations.—Continued

[Locations of source areas are shown in figure 3. Orphans are identified as particles that did not terminate in one of the specified source areas. None of the source areas include model layer 6. Streamflow reaches are located between the model boundary and streamflow-gaging station (map No.) 504 (600–601), within the Big Lost River spreading area (602–605), between streamflow-gaging stations 504 and 506 (606–607), within the Big Lost River sinks and playas (608–610), within the Little Lost River downstream of the model boundary (611), and within Birch Creek downstream of the model boundary (612). **FRAC:** the fraction of the total outflow from a cell contributed by an internal source; used to identify weak-source cells. **Layer:** the model layer particles are located in at their release. **Depth interval:** the depths of released particles are distributed randomly in this depth interval, in feet below the top of model layer 1. **Radius:** the aerial radius of the cylindrical volume defining particle release locations, in feet. **Abbreviation:** NA, not applicable]

Subregional layered-grid simulation; FRAC = 0.1 and Layer = 1–6

Simulated groundwater source area	Number of particles from source area within model layer						Percentage of source area contributions within model layer					
	1	2	3	4	5	1–6	1	2	3	4	5	1–6
Northwest mountain-front boundary												
Big Lost River valley (BLR)	4,867	NA	NA	NA	NA	4,867	29.1	NA	NA	NA	NA	11.6
Little Lost River valley (LLR)	1,254	1,467	1,624	NA	NA	4,345	7.5	24.8	30.3	NA	NA	10.3
Birch Creek valley (BC)	628	684	470	853	NA	2,635	3.8	11.6	8.8	10.5	NA	6.3
Northwest mountain-front subtotal	6,749	2,151	2,094	853	NA	11,847	40.4	36.4	39.1	10.5	NA	28.2
Northeast regional-underflow boundary												
Reno Ranch section (Re)	297	274	430	553	656	2,210	1.8	4.6	8.0	6.8	11.0	5.3
Monteview section (Mo)	652	339	352	813	NA	2,156	3.9	5.7	6.6	10.0	NA	5.1
Mud Lake section (ML)	1,999	1,383	1,136	3,075	NA	7,593	12.0	23.4	21.2	37.8	NA	18.1
Terreton section (Te)	2,046	1,698	1,344	2,833	5,296	13,217	12.2	28.7	25.1	34.9	89.0	31.4
Northeast regional-underflow subtotal	4,994	3,694	3,262	7,274	5,952	25,176	29.9	62.5	60.9	89.5	100.0	59.9
Water table boundary												
Streamflow infiltration												
Big Lost River infiltration												
Stream reach 600–601	903	NA	NA	NA	NA	903	5.4	NA	NA	NA	NA	2.1
Stream reach 602–605	659	NA	NA	NA	NA	659	3.9	NA	NA	NA	NA	1.6
Stream reach 606–607	553	NA	NA	NA	NA	553	3.3	NA	NA	NA	NA	1.3
Stream reach 608–610	1,451	NA	NA	NA	NA	1,451	8.7	NA	NA	NA	NA	3.5
Little Lost River infiltration												
Stream reach 611	90	NA	NA	NA	NA	90	0.5	NA	NA	NA	NA	0.2
Birch Creek infiltration												
Stream reach 612	856	NA	NA	NA	NA	856	5.1	NA	NA	NA	NA	2.0
Streamflow-infiltration subtotal	4,512	NA	NA	NA	NA	4,512	27.0	NA	NA	NA	NA	10.7
Orphans	456	62	1	0	1	520	2.7	1.0	0.0	0.0	0.0	1.2
Total	16,711	5,907	5,357	8,127	5,953	42,055	100.0	100.0	100.0	100.0	100.0	100.0

Appendix G. Summary of backward particle-tracking results from a sensitivity analysis of the source area contributions to changes in input parameters for the layered-grid and NPR-W01 well simulations.—Continued

[Locations of source areas are shown in figure 3. Orphans are identified as particles that did not terminate in one of the specified source areas. None of the source areas include model layer 6. Streamflow reaches are located between the model boundary and streamflow-gaging station (map No., 504 (600–601), within the Big Lost River spreading area (602–605), between streamflow-gaging stations 504 and 506 (606–607), within the Big Lost River sinks and playas (608–610), within the Little Lost River downstream of the model boundary (611), and within Birch Creek downstream of the model boundary (612). **FRAC:** the fraction of the total outflow from a cell contributed by an internal source; used to identify weak-source cells. **Layer:** the model layer particles are located in at their release. **Depth interval:** the depths of released particles are distributed randomly in this depth interval, in feet below the top of model layer 1. **Radius:** the aerial radius of the cylindrical volume defining particle release locations, in feet. **Abbreviation:** NA, not applicable]

Subregional layered-grid simulation; FRAC = 0.9 and Layer = 1-6

Simulated groundwater source area	Number of particles from source area within model layer						Percentage of source area contributions within model layer					
	1	2	3	4	5	1–6	1	2	3	4	5	1–6
Northwest mountain-front boundary												
Big Lost River valley (BLR)	5,303	NA	NA	NA	NA	5,303	32.2	NA	NA	NA	NA	12.6
Little Lost River valley (LLR)	3,478	1,750	1,626	NA	NA	6,854	21.1	28.5	30.3	NA	NA	16.3
Birch Creek valley (BC)	727	684	470	853	NA	2,734	4.4	11.2	8.8	10.5	NA	6.5
Northwest mountain-front subtotal	9,508	2,434	2,096	853	NA	14,891	57.7	39.7	39.1	10.5		35.4
Northeast regional-underflow boundary												
Reno Ranch section (Re)	297	274	430	554	656	2,211	1.8	4.5	8.0	6.8	11.0	5.3
Monteview section (Mo)	652	339	352	813	NA	2,156	4.0	5.5	6.5	10.0	NA	5.1
Mud Lake sector (ML)	1,999	1,388	1,137	3,075	NA	7,599	12.1	22.6	21.2	37.8	NA	18.1
Terreton section (Te)	2,046	1,698	1,344	2,833	5,296	13,217	12.4	27.7	25.1	34.9	89.0	31.4
Northeast regional-underflow subtotal	4,994	3,699	3,263	7,275	5,952	25,183	30.3	60.3	60.9	89.5	100.0	59.9
Water table boundary												
Streamflow infiltration												
Big Lost River infiltration												
Stream reach 600–601	0	NA	NA	NA	NA	0	0.0	NA	NA	NA	NA	0.0
Stream reach 602–605	0	NA	NA	NA	NA	0	0.0	NA	NA	NA	NA	0.0
Stream reach 606–607	206	NA	NA	NA	NA	206	1.2	NA	NA	NA	NA	0.5
Stream reach 608–610	190	NA	NA	NA	NA	190	1.2	NA	NA	NA	NA	0.5
Little Lost River infiltration												
Stream reach 611	0	NA	NA	NA	NA	0	0.0	NA	NA	NA	NA	0.0
Birch Creek infiltration												
Stream reach 612	942	NA	NA	NA	NA	942	5.7	NA	NA	NA	NA	2.2
Streamflow-infiltration subtotal	1,338	NA	NA	NA	NA	1,338	8.1	NA	NA	NA	NA	3.2
Orphans	641	0	1	0	1	643	3.9	0.0	0.0	0.0	0.0	1.5
Total	16,481	6,133	5,360	8,128	5,953	42,055	100.0	100.0	100.0	100.0	100.0	100.0

Appendix G. Summary of backward particle-tracking results from a sensitivity analysis of the source area contributions to changes in input parameters for the layered-grid and NPR-W01 well simulations.—Continued

[Locations of source areas are shown in figure 3. Orphans are identified as particles that did not terminate in one of the specified source areas. None of the source areas include model layer 6. Streamflow reaches are located between the model boundary and streamflow-gaging station (map No.) 504 (600–601), within the Big Lost River spreading area (602–605), between streamflow-gaging stations 504 and 506 (606–607), within the Big Lost River sinks and playas (608–610), within the Little Lost River downstream of the model boundary (611), and within Birch Creek downstream of the model boundary (612). **FRAC:** the fraction of the total outflow from a cell contributed by an internal source; used to identify weak-source cells. **Layer:** the model layer particles are located in at their release. **Depth interval:** the depths of released particles are distributed randomly in this depth interval, in feet below the top of model layer 1. **Radius:** the aerial radius of the cylindrical volume defining particle release locations, in feet. **Abbreviation:** NA, not applicable]

Subregional layered-grid simulation; FRAC= 0.5 and Layer = 1

Simulated groundwater source area	Number of particles from source area within model layer						Percentage of source area contributions within model layer					
	1	2	3	4	5	1–6	1	2	3	4	5	1–6
Northwest mountain-front boundary												
Big Lost River valley (BLR)	1,199	NA	NA	NA	NA	1,199	17.6	NA	NA	NA	NA	15.6
Little Lost River valley (LLR)	1,362	48	0	NA	NA	1,410	20.0	6.6	0.0	NA	NA	18.3
Birch Creek valley (BC)	200	0	0	0	NA	200	2.9	0.0	0.0	0.0	NA	2.6
Northwest mountain-front subtotal	2,761	48	0	0	NA	2,809	40.6	6.6	0.0	0.0	NA	36.5
Northeast regional-underflow boundary												
Reno Ranch section (Re)	117	0	4	4	0	125	1.7	0.0	4.2	5.9	0.0	1.6
Monteview section (Mo)	188	0	0	0	NA	188	2.8	0.0	0.0	0.0	NA	2.4
Mud Lake section (ML)	1,167	201	2	0	NA	1,370	17.2	27.7	2.1	0.0	NA	17.8
Terreton section (Te)	1,092	477	90	64	11	1,734	16.1	65.7	93.8	94.1	100.0	22.5
Northeast regional-underflow subtotal	2,564	678	96	68	11	3,417	37.7	93.4	100.0	100.0	100.0	44.4
Water table boundary												
Streamflow infiltration												
Big Lost River infiltration												
Stream reach 600–601	15	NA	NA	NA	NA	15	0.2	NA	NA	NA	NA	0.2
Stream reach 602–605	0	NA	NA	NA	NA	0	0.0	NA	NA	NA	NA	0.0
Stream reach 606–607	302	NA	NA	NA	NA	302	4.4	NA	NA	NA	NA	3.9
Stream reach 608–610	461	NA	NA	NA	NA	461	6.8	NA	NA	NA	NA	6.0
Little Lost River infiltration												
Stream reach 611	0	NA	NA	NA	NA	0	0.0	NA	NA	NA	NA	0.0
Birch Creek infiltration												
Stream reach 612	270	NA	NA	NA	NA	270	4.0	NA	NA	NA	NA	3.5
Streamflow-infiltration subtotal	1,048	NA	NA	NA	NA	1,048	15.4	NA	NA	NA	NA	13.6
Orphans	426	0	0	0	0	426	6.3	0.0	0.0	0.0	0.0	5.5
Total	6,799	726	96	68	11	7,700	100.0	100.0	100.0	100.0	100.0	100.0

Appendix G. Summary of backward particle-tracking results from a sensitivity analysis of the source area contributions to changes in input parameters for the layered-grid and NPR-W01 well simulations.—Continued

[Locations of source areas are shown in figure 3. Orphans are identified as particles that did not terminate in one of the specified source areas. None of the source areas include model layer 6. Streamflow reaches are located between the model boundary and streamflow-gaging station (map No.) 504 (600–601), within the Big Lost River spreading area (602–605), between streamflow-gaging stations 504 and 506 (606–607), within the Big Lost River sinks and playas (608–610), within the Little Lost River downstream of the model boundary (611), and within Birch Creek downstream of the model boundary (612). **FRAC:** the fraction of the total outflow from a cell contributed by an internal source; used to identify weak-source cells. **Layer:** the model layer particles are located in at their release. **Depth interval:** the depths of released particles are distributed randomly in this depth interval, in feet below the top of model layer 1. **Radius:** the aerial radius of the cylindrical volume defining particle release locations, in feet. **Abbreviation:** NA, not applicable]

Subregional layered-grid simulation; FRAC = 0.5 and Layer = 2

Simulated groundwater source area	Number of particles from source area within model layer						Percentage of source area contributions within model layer					
	1	2	3	4	5	1–6	1	2	3	4	5	1–6
Northwest mountain-front boundary												
Big Lost River valley (BLR)	1,272	NA	NA	NA	NA	1,272	30.2	NA	NA	NA	NA	16.5
Little Lost River valley (LLR)	630	706	207	NA	NA	1,543	15.0	40.8	26.8	NA	NA	20.0
Birch Creek valley (BC)	132	156	0	0	NA	288	3.1	9.0	0.0	0.0	NA	3.7
Northwest mountain-front subtotal	2,034	862	207	0	NA	3,103	48.3	49.9	26.8	0.0	NA	40.3
Northeast regional-underflow boundary												
Reno Ranch section (Re)	69	132	2	2	0	205	1.6	7.6	0.3	0.2	0.0	2.7
Monteview section (Mo)	207	47	0	0	NA	254	4.9	2.7	0.0	0.0	NA	3.3
Mud Lake section (ML)	490	412	196	208	NA	1,306	11.6	23.8	25.4	21.1	NA	16.9
Terreton section (Te)	590	276	368	774	10	2,018	14.0	16.0	47.5	78.7	100.0	26.2
Northeast regional-underflow subtotal	1,356	867	566	984	10	3,783	32.2	50.1	73.2	100.0	100.0	49.1
Water table boundary												
Streamflow infiltration												
Big Lost River infiltration												
Stream reach 600–601	13	NA	NA	NA	NA	13	0.3	NA	NA	NA	NA	0.2
Stream reach 602–605	0	NA	NA	NA	NA	0	0.0	NA	NA	NA	NA	0.0
Stream reach 606–607	8	NA	NA	NA	NA	8	0.2	NA	NA	NA	NA	0.1
Stream reach 608–610	442	NA	NA	NA	NA	442	10.5	NA	NA	NA	NA	5.7
Little Lost River infiltration												
Stream reach 611	0	NA	NA	NA	NA	0	0.0	NA	NA	NA	NA	0.0
Birch Creek infiltration												
Stream reach 612	294	NA	NA	NA	NA	294	7.0	NA	NA	NA	NA	3.8
Streamflow-infiltration subtotal	757	NA	NA	NA	NA	757	18.0	NA	NA	NA	NA	9.8
Orphans	65	0	0	0	0	65	1.5	0.0	0.0	0.0	0.0	0.8
Total	4,212	1,729	773	984	10	7,708	100.0	100.0	100.0	100.0	100.0	100.0

Appendix G. Summary of backward particle-tracking results from a sensitivity analysis of the source area contributions to changes in input parameters for the layered-grid and NPR-W01 well simulations.—Continued

[Locations of source areas are shown in figure 3. Orphans are identified as particles that did not terminate in one of the specified source areas. None of the source areas include model layer 6. Streamflow reaches are located between the model boundary and streamflow-gaging station (map No.) 504 (600–601), within the Big Lost River spreading area (602–605), between streamflow-gaging stations 504 and 506 (606–607), within the Big Lost River sinks and playas (608–610), within the Little Lost River downstream of the model boundary (611), and within Birch Creek downstream of the model boundary (612). **FRAC:** the fraction of the total outflow from a cell contributed by an internal source; used to identify weak-source cells. **Layer:** the model layer particles are located in at their release. **Depth interval:** the depths of released particles are distributed randomly in this depth interval, in feet below the top of model layer 1. **Radius:** the aerial radius of the cylindrical volume defining particle release locations, in feet. **Abbreviation:** NA, not applicable]

Subregional layered-grid simulation; FRAC = 0.5 and Layer = 3

Simulated groundwater source area	Number of particles from source area within model layer						Percentage of source area contributions within model layer					
	1	2	3	4	5	1-6	1	2	3	4	5	1-6
Northwest mountain-front boundary												
Big Lost River valley (BLR)	1,161	NA	NA	NA	NA	1,161	40.6	NA	NA	NA	NA	15.3
Little Lost River valley (LLR)	353	561	569	NA	NA	1,483	12.3	38.9	39.2	NA	NA	19.6
Birch Creek valley (BC)	70	146	180	0	NA	396	2.4	10.1	12.4	0.0	NA	5.2
Northwest mountain-front subtotal	1,584	707	749	0	NA	3,040	55.4	49.0	51.7	0.0	NA	40.2
Northeast regional-underflow boundary												
Reno Ranch section (Re)	57	22	153	2	0	234	2.0	1.5	10.6	0.1	0.0	3.1
Monteview section (Mo)	160	35	42	0	NA	237	5.6	2.4	2.9	0.0	NA	3.1
Mud Lake section (ML)	278	334	338	696	NA	1,646	9.7	23.1	23.3	46.8	NA	21.7
Terreton section (Te)	320	345	168	789	332	1,954	11.2	23.9	11.6	53.1	100.0	25.8
Northeast regional-underflow subtotal	815	736	701	1,487	332	4,071	28.5	51.0	48.3	100.0	100.0	53.8
Water table boundary												
Streamflow infiltration												
Big Lost River infiltration												
Stream reach 600–601	10	NA	NA	NA	NA	10	0.3	NA	NA	NA	NA	0.1
Stream reach 602–605	0	NA	NA	NA	NA	0	0.0	NA	NA	NA	NA	0.0
Stream reach 606–607	2	NA	NA	NA	NA	2	0.1	NA	NA	NA	NA	0.0
Stream reach 608–610	225	NA	NA	NA	NA	225	7.9	NA	NA	NA	NA	3.0
Little Lost River infiltration												
Stream reach 611	0	NA	NA	NA	NA	0	0.0	NA	NA	NA	NA	0.0
Birch Creek infiltration												
Stream reach 612	199	NA	NA	NA	NA	199	7.0	NA	NA	NA	NA	2.6
Streamflow-infiltration subtotal	436	NA	NA	NA	NA	436	15.3	NA	NA	NA	NA	5.8
Orphans	24	0	0	0	0	24	0.8	0.0	0.0	0.0	0.0	0.3
Total	2,859	1,443	1,450	1,487	332	7,571	100.0	100.0	100.0	100.0	100.0	100.0

Appendix G. Summary of backward particle-tracking results from a sensitivity analysis of the source area contributions to changes in input parameters for the layered-grid and NPR-W01 well simulations.—Continued

[Locations of source areas are shown in figure 3. Orphans are identified as particles that did not terminate in one of the specified source areas. None of the source areas include model layer 6. Streamflow reaches are located between the model boundary and streamflow-gaging station (map No.) 504 (600–601), within the Big Lost River spreading area (602–605), between streamflow-gaging stations 504 and 506 (606–607), within the Big Lost River sinks and playas (608–610), within the Little Lost River downstream of the model boundary (611), and within Birch Creek downstream of the model boundary (612). **FRAC:** the fraction of the total outflow from a cell contributed by an internal source; used to identify weak-source cells. **Layer:** the model layer particles are located in at their release. **Radius:** the aerial radius of the cylindrical volume defining particle release locations, in feet. **Depth interval:** the depths of released particles are distributed randomly in this depth interval, in feet below the top of model layer 1. **Abbreviation:** NA, not applicable]

Subregional layered-grid simulation; FRAC = 0.5 and Layer = 4

Simulated groundwater source area	Number of particles from source area within model layer						Percentage of source area contributions within model layer					
	1	2	3	4	5	1–6	1	2	3	4	5	1–6
Northwest mountain-front boundary												
Big Lost River valley (BLR)	1,106	NA	NA	NA	NA	1,106	67.1	NA	NA	NA	NA	14.9
Little Lost River valley (LLR)	122	317	426	NA	NA	865	7.4	18.9	41.9	NA	NA	11.6
Birch Creek valley (BC)	74	182	141	370	NA	767	4.5	10.9	13.9	15.9	NA	10.3
Northwest mountain-front subtotal	1,302	499	567	370	NA	2,738	79.0	29.8	55.8	15.9		36.8
Northeast regional-underflow boundary												
Reno Ranch section (Re)	38	77	12	181	0	308	2.3	4.6	1.2	7.8	0.0	4.1
Monteview section (Mo)	53	201	31	67	NA	352	3.2	12.0	3.0	2.9	NA	4.7
Mud Lake section (ML)	43	391	246	1,119	NA	1,799	2.6	23.3	24.2	48.0	NA	24.2
Terreton section (Te)	36	508	161	595	758	2,058	2.2	30.3	15.8	25.5	99.9	27.7
Northeast regional-underflow subtotal	170	1177	450	1,962	758	4,517	10.3	70.2	44.2	84.1	99.9	60.8
Water table boundary												
Streamflow infiltration												
Big Lost River infiltration												
Stream reach 600–601	9	NA	NA	NA	NA	9	0.5	NA	NA	NA	NA	0.1
Stream reach 602–605	0	NA	NA	NA	NA	0	0.0	NA	NA	NA	NA	0.0
Stream reach 606–607	5	NA	NA	NA	NA	5	0.3	NA	NA	NA	NA	0.1
Stream reach 608–610	48	NA	NA	NA	NA	48	2.9	NA	NA	NA	NA	0.6
Little Lost River infiltration												
Stream reach 611	0	NA	NA	NA	NA	0	0.0	NA	NA	NA	NA	0.0
Birch Creek infiltration												
Stream reach 612	107	NA	NA	NA	NA	107	6.5	NA	NA	NA	NA	1.4
Streamflow-infiltration subtotal	169	NA	NA	NA	NA	169	10.3	NA	NA	NA	NA	2.3
Orphans	7	0	0	0	1	8	0.4	0.0	0.0	0.0	0.1	0.1
Total	1,648	1,676	1,017	2,332	759	7,432	100.0	100.0	100.0	100.0	100.0	100.0

Appendix G. Summary of backward particle-tracking results from a sensitivity analysis of the source area contributions to changes in input parameters for the layered-grid and NPR-W01 well simulations.—Continued

[Locations of source areas are shown in figure 3. Orphans are identified as particles that did not terminate in one of the specified source areas. None of the source areas include model layer 6. Streamflow reaches are located between the model boundary and streamflow-gaging station (map No.) 504 (600–601), within the Big Lost River spreading area (602–605), between streamflow-gaging stations 504 and 506 (606–607), within the Big Lost River sinks and playas (608–610), within the Little Lost River downstream of the model boundary (611), and within Birch Creek downstream of the model boundary (612). **FRAC:** the fraction of the total outflow from a cell contributed by an internal source; used to identify weak-source cells. **Layer:** the model layer particles are located in at their release. **Depth interval:** the depths of released particles are distributed randomly in this depth interval, in feet below the top of model layer 1. **Radius:** the aerial radius of the cylindrical volume defining particle release locations, in feet. **Abbreviation:** NA, not applicable]

Subregional layered-grid simulation; FRAC = 0.5 and Layer = 5

Simulated groundwater source area	Number of particles from source area within model layer						Percentage of source area contributions within model layer					
	1	2	3	4	5	1–6	1	2	3	4	5	1–6
Northwest mountain-front boundary												
Big Lost River valley (BLR)	438	NA	NA	NA	NA	438	65.6	NA	NA	NA	NA	7.2
Little Lost River valley (LLR)	24	65	285	NA	NA	374	3.6	20.8	18.9	NA	NA	6.1
Birch Creek valley (BC)	32	32	36	273	NA	373	4.8	10.3	2.4	16.0	NA	6.1
Northwest mountain-front subtotal	494	97	321	273	NA	1,185	74.0	31.1	21.3	16.0	NA	19.4
Northeast regional-underflow boundary												
Reno Ranch section (Re)	16	18	143	69	248	494	2.4	5.8	9.5	4.0	13.0	8.1
Monteview section (Mo)	44	55	133	253	NA	485	6.6	17.6	8.8	14.8	NA	7.9
Mud Lake section (ML)	21	50	355	602	NA	1,028	3.1	16.0	23.5	35.3	NA	16.8
Terreton section (Te)	8	92	557	508	1,660	2,825	1.2	29.5	36.9	29.8	87.0	46.3
Northeast regional-underflow subtotal	89	215	1,188	1,432	1,908	4,832	13.3	68.9	78.7	84.0	100.0	79.2
Water table boundary												
Streamflow infiltration												
Big Lost River infiltration												
Stream reach 600–601	0	NA	NA	NA	NA	0	0.0	NA	NA	NA	NA	0.0
Stream reach 602–605	0	NA	NA	NA	NA	0	0.0	NA	NA	NA	NA	0.0
Stream reach 606–607	0	NA	NA	NA	NA	0	0.0	NA	NA	NA	NA	0.0
Stream reach 608–610	18	NA	NA	NA	NA	18	2.7	NA	NA	NA	NA	0.3
Little Lost River infiltration												
Stream reach 611	0	NA	NA	NA	NA	0	0.0	NA	NA	NA	NA	0.0
Birch Creek infiltration												
Stream reach 612	49	NA	NA	NA	NA	49	7.3	NA	NA	NA	NA	0.8
Streamflow-infiltration subtotal	67	NA	NA	NA	NA	67	10.0	NA	NA	NA	NA	1.1
Orphans	18	0	1	0	0	19	2.7	0.0	0.1	0.0	0.0	0.3
Total	668	312	1,510	1,705	1,908	6,103	100.0	100.0	100.0	100.0	100.0	100.0

Appendix G. Summary of backward particle-tracking results from a sensitivity analysis of the source area contributions to changes in input parameters for the layered-grid and NPR-W01 well simulations.—Continued

[Locations of source areas are shown in figure 3. Orphans are identified as particles that did not terminate in one of the specified source areas. None of the source areas include model layer 6. Streamflow reaches are located between the model boundary and streamflow-gaging station (map No.; 504 (600–601), within the Big Lost River spreading area (602–605), between streamflow-gaging stations 504 and 506 (606–607), within the Big Lost River sinks and playas (608–610), within the Little Lost River downstream of the model boundary (611), and within Birch Creek downstream of the model boundary (612). **FRAC:** the fraction of the total outflow from a cell contributed by an internal source; used to identify weak-source cells. **Layer:** the model layer particles are located in at their release. **Depth interval:** the depths of released particles are distributed randomly in this depth interval, in feet below the top of model layer 1. **Radius:** the aerial radius of the cylindrical volume defining particle release locations, in feet. **Abbreviation:** NA, not applicable]

Subregional layered-grid simulation; FRAC = 0.5 and Layer = 6

Simulated groundwater source area	Number of particles from source area within model layer						Percentage of source area contributions within model layer					
	1	2	3	4	5	1–6	1	2	3	4	5	1–6
Northwest mountain-front boundary												
Big Lost River valley (BLR)	127	NA	NA	NA	NA	127	42.9	NA	NA	NA	NA	2.3
Little Lost River valley (LLR)	21	52	139	NA	NA	212	7.1	NA	NA	NA	NA	3.8
Birch Creek valley (BC)	142	168	113	210	NA	633	48.0	NA	NA	NA	NA	11.4
Northwest mountain-front subtotal	290	220	252	210	NA	972	98.0	NA	NA	NA	NA	17.5
Northeast regional-underflow boundary												
Reno Ranch section (Re)	0	25	116	296	408	845	0.0	NA	NA	NA	NA	15.2
Monteview section (Mo)	0	1	146	493	NA	640	0.0	NA	NA	NA	NA	11.6
Mud Lake section (ML)	0	0	0	450	NA	450	0.0	NA	NA	NA	NA	8.1
Terreton section (Te)	0	0	0	103	2,525	2,628	0.0	NA	NA	NA	NA	47.4
Northeast regional-underflow subtotal	0	26	262	1,342	2,933	4,563	0.0	NA	NA	NA	NA	82.3
Water table boundary												
Streamflow infiltration												
Big Lost River infiltration												
Stream reach 600–601	0	NA	NA	NA	NA	0	0.0	NA	NA	NA	NA	0.0
Stream reach 602–605	0	NA	NA	NA	NA	0	0.0	NA	NA	NA	NA	0.0
Stream reach 606–607	0	NA	NA	NA	NA	0	0.0	NA	NA	NA	NA	0.0
Stream reach 608–610	0	NA	NA	NA	NA	0	0.0	NA	NA	NA	NA	0.0
Little Lost River infiltration												
Stream reach 611	0	NA	NA	NA	NA	0	0.0	NA	NA	NA	NA	0.0
Birch Creek infiltration												
Stream reach 612	4	NA	NA	NA	NA	4	1.4	NA	NA	NA	NA	0.1
Streamflow-infiltration subtotal	4	NA	NA	NA	NA	4	1.4	NA	NA	NA	NA	0.1
Orphans	2	0	0	0	0	2	0.7	NA	NA	NA	NA	0.0
Total	296	246	514	1,552	2,933	5,541	100.0	NA	NA	NA	NA	100.0

Appendix G.   Summary of backward particle-tracking results from a sensitivity analysis of the source area contributions to changes in input parameters for the layered-grid and NPR-W01 well simulations.—Continued

[Locations of source areas are shown in figure 3. Orphans are identified as particles that did not terminate in one of the specified source areas. None of the source areas include model layer 6. Streamflow reaches are located between the model boundary and streamflow-gaging station (map No.) 504 (600–601), within the Big Lost River spreading area (602–605), between streamflow-gaging stations 504 and 506 (606–607), within the Big Lost River sinks and playas (608–610), within the Little Lost River downstream of the model boundary (611), and within Birch Creek downstream of the model boundary (612). **FRAC:** the fraction of the total outflow from a cell contributed by an internal source; used to identify weak-source cells. **Layer:** the model layer particles are located in at their release. **Depth interval:** the depths of released particles are distributed randomly in this depth interval, in feet below the top of model layer 1. **Radius:** the aerial radius of the cylindrical volume defining particle release locations, in feet. **Abbreviation:** NA, not applicable]

NPR-W01 well simulation; FRAC = 0.5, Depth interval = 0–200 feet, and Radius = 2,000 feet (base case)

Simulated groundwater source area	Number of particles from source area within model layer						Percentage of source area contributions within model layer					
	1	2	3	4	5	1–6	1	2	3	4	5	1–6
Northwest mountain-front boundary												
Big Lost River valley (BLR)	0	NA	NA	NA	NA	0	0.0	NA	NA	NA	NA	0.0
Little Lost River valley (LLR)	321	0	0	NA	NA	321	4.5	0.0	0.0	NA	NA	3.6
Birch Creek valley (BC)	0	0	0	0	NA	0	0.0	0.0	0.0	0.0	NA	0.0
Northwest mountain-front subtotal	321	0	0	0	0	321	4.5	0.0	0.0	0.0	0.0	3.6
Northeast regional-underflow boundary												
Reno Ranch section (Re)	9	0	0	0	0	9	0.1	0.0	0.0	0.0	0.0	0.1
Monteview section (Mo)	861	0	0	0	NA	861	12.1	0.0	0.0	0.0	NA	9.5
Mud Lake section (ML)	431	58	1,109	724	NA	2,322	6.0	100.0	100.0	100.0	NA	25.7
Terreton section (Te)	0	0	0	0	0	0	0.0	0.0	0.0	0.0	0.0	0.0
Northeast regional-underflow subtotal	1,301	58	1,109	724	0	3,192	18.2	100.0	100.0	100.0	0.0	35.4
Water table boundary												
Streamflow infiltration												
Big Lost River infiltration												
Stream reach 600–601	0	NA	NA	NA	NA	0	0.0	NA	NA	NA	NA	0.0
Stream reach 602–605	0	NA	NA	NA	NA	0	0.0	NA	NA	NA	NA	0.0
Stream reach 606–607	1,297	NA	NA	NA	NA	1,297	18.2	NA	NA	NA	NA	14.4
Stream reach 608–610	3,108	NA	NA	NA	NA	3,108	43.5	NA	NA	NA	NA	34.4
Little Lost River infiltration												
Stream reach 611	0	NA	NA	NA	NA	0	0.0	NA	NA	NA	NA	0.0
Birch Creek infiltration												
Stream reach 612	58	NA	NA	NA	NA	58	0.8	NA	NA	NA	NA	0.6
Streamflow-infiltration subtotal	4,463	NA	NA	NA	NA	4,463	62.5	NA	NA	NA	NA	49.4
Orphans	1,052	0	0	0	0	1,052	14.7	0.0	0.0	0.0	0.0	11.7
Total	7,137	58	1,109	724	0	9,028	100.0	100.0	100.0	100.0	0.0	100.0

Appendix G. Summary of backward particle-tracking results from a sensitivity analysis of the source area contributions to changes in input parameters for the layered-grid and NPR-W01 well simulations.—Continued

[Locations of source areas are shown in figure 3. Orphans are identified as particles that did not terminate in one of the specified source areas. None of the source areas include model layer 6. Streamflow reaches are located between the model boundary and streamflow-gaging station (map No.) 504 (600–601), within the Big Lost River spreading area (602–605), between streamflow-gaging stations 504 and 506 (606–607), within the Big Lost River sinks and playas (608–610), within the Little Lost River downstream of the model boundary (611), and within Birch Creek downstream of the model boundary (612). **FRAC:** the fraction of the total outflow from a cell contributed by an internal source; used to identify weak-source cells. **Layer:** the model layer particles are located in at their release. **Radius:** the aerial radius of the cylindrical volume defining particle release locations, in feet. **Depth interval:** the depths of released particles are distributed randomly in this depth interval, in feet below the top of model layer 1. Abbreviation: NA, not applicable]

NPR-W01 well simulation; FRAC = 0.1, Depth interval = 0–200 feet, and Radius = 2,000 feet

Simulated groundwater source area	Number of particles from source area within model layer						Percentage of source area contributions within model layer					
	1	2	3	4	5	1–6	1	2	3	4	5	1–6
Northwest mountain-front boundary												
Big Lost River valley (BLR)	0	NA	NA	NA	NA	0	0.0	NA	NA	NA	NA	0.0
Little Lost River valley (LLR)	0	0	0	NA	NA	0	0.0	0.0	0.0	NA	NA	0.0
Birch Creek valley (BC)	0	0	0	0	NA	0	0.0	0.0	0.0	0.0	NA	0.0
Northwest mountain-front subtotal	0	0	0	0	NA	0	0.0	0.0	0.0	0.0	NA	0.0
Northeast regional-underflow boundary												
Reno Ranch section (Re)	9	0	0	0	0	9	0.1	0.0	0.0	0.0	0.0	0.1
Monteview section (Mo)	861	0	0	0	NA	861	12.1	0.0	0.0	0.0	NA	9.5
Mud Lake section (ML)	431	58	1,109	724	NA	2,322	6.0	100.0	100.0	100.0	NA	25.7
Terreton section (Te)	0	0	0	0	0	0	0.0	0.0	0.0	0.0	0.0	0.0
Northeast regional-underflow subtotal	1,301	58	1,109	724	0	3,192	18.2	100.0	100.0	100.0	0.0	35.4
Water table boundary												
Streamflow infiltration												
Big Lost River infiltration												
Stream reach 600–601	0	NA	NA	NA	NA	0	0.0	NA	NA	NA	NA	0.0
Stream reach 602–605	0	NA	NA	NA	NA	0	0.0	NA	NA	NA	NA	0.0
Stream reach 606–607	3,113	NA	NA	NA	NA	3,113	43.6	NA	NA	NA	NA	34.5
Stream reach 608–610	1,892	NA	NA	NA	NA	1,892	26.5	NA	NA	NA	NA	21.0
Little Lost River infiltration												
Stream reach 611	0	NA	NA	NA	NA	0	0.0	NA	NA	NA	NA	0.0
Birch Creek infiltration												
Stream reach 612	20	NA	NA	NA	NA	20	0.3	NA	NA	NA	NA	0.2
Streamflow-infiltration subtotal	5,025	NA	NA	NA	NA	5,025	70.4	NA	NA	NA	NA	55.7
Orphans	811	NA	0	0	0	811	11.4	0.0	0.0	0.0	0.0	9.0
Total	7,137	58	1,109	724	0	9,028	100.0	100.0	100.0	100.0	0.0	100.0

Appendix G. Summary of backward particle-tracking results from a sensitivity analysis of the source area contributions to changes in input parameters for the layered-grid and NPR-W01 well simulations.—Continued

[Locations of source areas are shown in figure 3. Orphans are identified as particles that did not terminate in one of the specified source areas. None of the source areas include model layer 6. Streamflow reaches are located between the model boundary and streamflow-gaging station (map No.) 504 (600–601), within the Big Lost River spreading area (602–605), between streamflow-gaging stations 504 and 506 (606–607), within the Big Lost River sinks and playas (608–610), within the Little Lost River downstream of the model boundary (611), and within Birch Creek downstream of the model boundary (612). **FRAC**: the fraction of the total outflow from a cell contributed by an internal source; used to identify weak-source cells. **Layer**: the model layer particles are located in at their release. **Depth interval**: the depths of released particles are distributed randomly in this depth interval, in feet below the top of model layer 1. **Radius**: the aerial radius of the cylindrical volume defining particle release locations, in feet. **Abbreviation**: NA, not applicable]

NPR-W01 well simulation; FRAC = 0.9, Depth interval = 0–200 feet, and Radius = 2,000 feet

Simulated groundwater source area	Number of particles from source area within model layer						Percentage of source area contributions within model layer					
	1	2	3	4	5	1–6	1	2	3	4	5	1–6
Northwest mountain-front boundary												
Big Lost River valley (BLR)	0	NA	NA	NA	NA	0	0.0	NA	NA	NA	NA	0.0
Little Lost River valley (LLR)	3,365	0	0	NA	NA	3,365	47.1	0.0	0.0	NA	NA	37.3
Birch Creek valley (BC)	1	0	0	0	NA	1	0.0	0.0	0.0	0.0	NA	0.0
Northwest mountain-front subtotal	3,366	0	0	0	NA	3,366	47.2	0.0	0.0	0.0	NA	37.3
Northeast regional-underflow boundary												
Reno Ranch section (Re)	9	0	0	0	0	9	0.1	0.0	0.0	0.0	0.0	0.1
Monteview section (Mo)	861	0	0	0	NA	861	12.1	0.0	0.0	0.0	NA	9.5
Mud Lake section (ML)	431	58	1,109	724	NA	2,322	6.0	100.0	100.0	100.0	NA	25.7
Terreton section (Te)	0	0	0	0	0	0	0.0	0.0	0.0	0.0	0.0	0.0
Northeast regional-underflow subtotal	1,301	58	1,109	724	0	3,192	18.2	100.0	100.0	100.0	0.0	35.4
Water table boundary												
Streamflow infiltration												
Big Lost River infiltration												
Stream reach 600–601	0	NA	NA	NA	NA	0	0.0	NA	NA	NA	NA	0.0
Stream reach 602–605	0	NA	NA	NA	NA	0	0.0	NA	NA	NA	NA	0.0
Stream reach 606–607	127	NA	NA	NA	NA	127	1.8	NA	NA	NA	NA	1.4
Stream reach 608–610	703	NA	NA	NA	NA	703	9.9	NA	NA	NA	NA	7.8
Little Lost River infiltration												
Stream reach 611	0	NA	NA	NA	NA	0	0.0	NA	NA	NA	NA	0.0
Birch Creek infiltration												
Stream reach 612	182	NA	NA	NA	NA	182	2.6	NA	NA	NA	NA	2.0
Streamflow-infiltration subtotal	1,012	NA	NA	NA	NA	1,012	14.2	NA	NA	NA	NA	11.2
Orphans	1,458	0	0	0	0	1,458	20.4	0.0	0.0	0.0	0.0	16.1
Total	7,137	58	1,109	724	0	9,028	100.0	100.0	100.0	100.0	0.0	100.0

Appendix G. Summary of backward particle-tracking results from a sensitivity analysis of the source area contributions to changes in input parameters for the layered-grid and NPR-W01 well simulations.—Continued

[Locations of source areas are shown in figure 3. Orphans are identified as particles that did not terminate in one of the specified source areas. None of the source areas include model layer 6. Streamflow reaches are located between the model boundary and streamflow-gaging station (map No.) 504 (600–601), within the Big Lost River spreading area (602–605), between streamflow-gaging stations 504 and 506 (606–607), within the Big Lost River sinks and playas (608–610), within the Little Lost River downstream of the model boundary (611), and within Birch Creek downstream of the model boundary (612). **FRAC:** the fraction of the total outflow from a cell contributed by an internal source; used to identify weak-source cells. **Layer:** the model layer particles are located in at their release. **Radius:** the aerial radius of the cylindrical volume defining particle release locations, in feet below the top of model layer 1. **Depth interval:** the depths of released particles are distributed randomly in this depth interval, in feet below the top of model layer 1. **Abbreviation:** NA, not applicable]

NPR-W01 well simulation; FRAC = 0.5, Depth interval = 0–50 feet, and Radius = 2,000 feet

Simulated groundwater source area	Number of particles from source area within model layer						Percentage of source area contributions within model layer					
	1	2	3	4	5	1–6	1	2	3	4	5	1–6
Northwest mountain-front boundary												
Big Lost River valley (BLR)	0	NA	NA	NA	NA	0	0.0	NA	NA	NA	NA	0.0
Little Lost River valley (LLR)	0	0	0	NA	NA	0	0.0	0.0	0.0	NA	NA	0.0
Birch Creek valley (BC)	0	0	0	0	NA	0	0.0	0.0	0.0	0.0	NA	0.0
Northwest mountain-front subtotal	0	0	0	0	NA	0	0.0	0.0	0.0	0.0	NA	0.0
Northeast regional-underflow boundary												
Reno Ranch section (Re)	0	0	0	0	0	0	0.0	0.0	0.0	0.0	0.0	0.0
Monteview section (Mo)	0	0	0	0	NA	0	0.0	0.0	0.0	0.0	NA	0.0
Mud Lake section (ML)	0	0	0	0	NA	0	0.0	0.0	0.0	0.0	NA	0.0
Terreton section (Te)	0	0	0	0	0	0	0.0	0.0	0.0	0.0	0.0	0.0
Northeast regional-underflow subtotal	0	0	0	0	0	0	0.0	0.0	0.0	0.0	0.0	0.0
Water table boundary												
Streamflow infiltration												
Big Lost River infiltration												
Stream reach 600–601	0	NA	NA	NA	NA	0	0.0	NA	NA	NA	NA	0.0
Stream reach 602–605	0	NA	NA	NA	NA	0	0.0	NA	NA	NA	NA	0.0
Stream reach 605–607	582	NA	NA	NA	NA	582	25.8	NA	NA	NA	NA	25.8
Stream reach 603–610	735	NA	NA	NA	NA	735	32.6	NA	NA	NA	NA	32.6
Little Lost River infiltration												
Stream reach 611	0	NA	NA	NA	NA	0	0.0	NA	NA	NA	NA	0.0
Birch Creek infiltration												
Stream reach 612	0	NA	NA	NA	NA	0	0.0	NA	NA	NA	NA	0.0
Streamflow-infiltration subtotal	1,317	NA	NA	NA	NA	1,317	58.4	NA	NA	NA	NA	58.4
Orphans	940	0	0	0	0	940	41.6	0.0	0.0	0.0	0.0	41.6
Total	2,257	0	0	0	0	2,257	100.0	0.0	0.0	0.0	0.0	100.0

Appendix G. Summary of backward particle-tracking results from a sensitivity analysis of the source area contributions to changes in input parameters for the layered-grid and NPR-W01 well simulations.—Continued

[Locations of source areas are shown in figure 3. Orphans are identified as particles that did not terminate in one of the specified source areas. None of the source areas include model layer 6. Streamflow reaches are located between the model boundary and streamflow-gaging station (map No.) 504 (600–601), within the Big Lost River spreading area (602–605), between streamflow-gaging stations 504 and 506 (606–607), within the Big Lost River sinks and playas (608–610), within the Little Lost River downstream of the model boundary (611), and within Birch Creek downstream of the model boundary (612). **FRAC:** the fraction of the total outflow from a cell contributed by an internal source; used to identify weak-source cells. **Layer:** the model layer particles are located in at their release. **Depth interval:** the depths of released particles are distributed randomly in this depth interval, in feet below the top of model layer 1. **Radius:** the aerial radius of the cylindrical volume defining particle release locations, in feet. **Abbreviation:** NA, not applicable]

NPR-W01 well simulation; FRAC = 0.5, Depth interval = 50–100 feet, and Radius = 2,000 feet

Simulated groundwater source area	Number of particles from source area within model layer						Percentage of source area contributions within model layer					
	1	2	3	4	5	1–6	1	2	3	4	5	1–6
Northwest mountain-front boundary												
Big Lost River valley (BLR)	0	NA	NA	NA	NA	0	0.0	NA	NA	NA	NA	0.0
Little Lost River valley (LLR)	2	0	NA	NA	NA	2	0.1	0.0	NA	NA	NA	0.1
Birch Creek valley (BC)	0	0	0	0	NA	0	0.0	0.0	0.0	0.0	NA	0.0
Northwest mountain-front subtotal	2	0	0	0	NA	2	0.1	0.0	0.0	0.0	NA	0.1
Northeast regional-underflow boundary												
Reno Ranch section (Re)	0	0	0	0	0	0	0.0	0.0	0.0	0.0	0.0	0.0
Monteview section (Mo)	0	0	0	0	NA	0	0.0	0.0	0.0	0.0	NA	0.0
Mud Lake section (ML)	0	0	0	0	NA	0	0.0	0.0	0.0	0.0	NA	0.0
Terreton section (Te)	0	0	0	0	0	0	0.0	0.0	0.0	0.0	0.0	0.0
Northeast regional-underflow subtotal	0	0	0	0	0	0	0.0	0.0	0.0	0.0	0.0	0.0
Water table boundary												
Streamflow infiltration												
Big Lost River infiltration												
Stream reach 600–601	0	NA	NA	NA	NA	0	0.0	NA	NA	NA	NA	0.0
Stream reach 602–605	0	NA	NA	NA	NA	0	0.0	NA	NA	NA	NA	0.0
Stream reach 606–607	699	NA	NA	NA	NA	699	31.0	NA	NA	NA	NA	31.0
Stream reach 608–610	1,554	NA	NA	NA	NA	1,554	68.9	NA	NA	NA	NA	68.9
Little Lost River infiltration												
Stream reach 611	0	NA	NA	NA	NA	0	0.0	NA	NA	NA	NA	0.0
Birch Creek infiltration												
Stream reach 612	0	NA	NA	NA	NA	0	0.0	NA	NA	NA	NA	0.0
Streamflow-infiltration subtotal	2,253	NA	NA	NA	NA	2,253	99.8	NA	NA	NA	NA	99.8
Orphans	2	0	0	0	0	2	0.1	0.0	0.0	0.0	0.0	0.1
Total	2,257	0	0	0	0	2,257	100.0	0.0	0.0	0.0	0.0	100.0

Appendix G. Summary of backward particle-tracking results from a sensitivity analysis of the source area contributions to changes in input parameters for the layered-grid and NPR-W01 well simulations.—Continued

[Locations of source areas are shown in figure 3. Orphans are identified as particles that did not terminate in one of the specified source areas. None of the source areas include model layer 6. Streamflow reaches are located between the model boundary and streamflow-gaging station (map No.) 504 (600–601), within the Big Lost River spreading area (602–605), between streamflow-gaging stations 504 and 506 (606–607), within the Big Lost River sinks and playas (608–610), within the Little Lost River downstream of the model boundary (611), and within Birch Creek downstream of the model boundary (612). **FRAC:** the fraction of the total outflow from a cell contributed by an internal source; used to identify weak-source cells. **Layer:** the model layer particles are located in at their release. **Depth interval:** the depths of released particles are distributed randomly in this depth interval, in feet below the top of model layer 1. **Radius:** the aerial radius of the cylindrical volume defining particle release locations, in feet. Abbreviation: NA, not applicable]

NPR-W01 well simulation; FRAC = 0.5, Depth interval = 100–150 feet, and Radius = 2,000 feet

Simulated groundwater source area	Number of particles from source area within model layer						Percentage of source area contributions within model layer					
	1	2	3	4	5	1–6	1	2	3	4	5	1–6
Northwest mountain-front boundary												
Big Lost River valley (BLR)	0	NA	NA	NA	NA	0	0.0	NA	NA	NA	NA	0.0
Little Lost River valley (LLR)	319	0	0	NA	NA	319	17.1	0.0	0.0	NA	NA	14.1
Birch Creek valley (BC)	0	0	0	0	NA	0	0.0	0.0	0.0	0.0	NA	0.0
Northwest mountain-front subtotal	319	0	0	0	NA	319	17.1	0.0	0.0	0.0	NA	14.1
Northeast regional-underflow boundary												
Reno Ranch section (Re)	9	0	0	0	0	9	0.5	0.0	0.0	0.0	0.0	0.4
Monteview section (Mo)	142	0	0	0	NA	142	7.6	0.0	0.0	0.0	NA	6.3
Mud Lake section (ML)	392	6	249	137	NA	784	21.0	100.0	100.0	100.0	NA	34.7
Terreton section (T)	0	0	0	0	0	0	0.0	0.0	0.0	0.0	0.0	0.0
Northeast regional-underflow subtotal	543	6	249	137	0	935	29.1	100.0	100.0	100.0	0.0	41.4
Water table boundary												
Streamflow infiltration												
Big Lost River infiltration												
Stream reach 600–601	0	NA	NA	NA	NA	0	0.0	NA	NA	NA	NA	0.0
Stream reach 602–605	0	NA	NA	NA	NA	0	0.0	NA	NA	NA	NA	0.0
Stream reach 606–607	16	NA	NA	NA	NA	16	0.9	NA	NA	NA	NA	0.7
Stream reach 608–610	819	NA	NA	NA	NA	819	43.9	NA	NA	NA	NA	36.3
Little Lost River infiltration												
Stream reach 611	0	NA	NA	NA	NA	0	0.0	NA	NA	NA	NA	0.0
Birch Creek infiltration												
Stream reach 612	58	NA	NA	NA	NA	58	3.1	NA	NA	NA	NA	2.6
Streamflow-infiltration subtotal	893	NA	NA	NA	NA	893	47.9	NA	NA	NA	NA	39.6
Orphans	110	0	0	0	0	110	5.9	0.0	0.0	0.0	0.0	4.9
Total	1,865	6	249	137	0	2,257	100.0	100.0	100.0	100.0	0.0	100.0

Appendix G. Summary of backward particle-tracking results from a sensitivity analysis of the source area contributions to changes in input parameters for the layered-grid and NPR-W01 well simulations.—Continued

[Locations of source areas are shown in figure 3. Orphans are identified as particles that did not terminate in one of the specified source areas. None of the source areas include model layer 6. Streamflow reaches are located between the model boundary and streamflow-gaging station (map No.) 504 (600–601), within the Big Lost River spreading area (602–605), between streamflow-gaging stations 504 and 506 (606–607), within the Big Lost River sinks and playas (608–610), within the Little Lost River downstream of the model boundary (611), and within Birch Creek downstream of the model boundary (612). FRAC: the fraction of the total outflow from a cell contributed by an internal source; used to identify weak-source cells. Layer: the model layer particles are located in at their release. Depth interval: the depths of released particles are distributed randomly in this depth interval, in feet below the top of model layer 1. Radius: the aerial radius of the cylindrical volume defining particle release locations, in feet. Abbreviation: NA, not applicable]

NPR-W01 well simulation; FRAC = 0.5, Depth interval = 150–200 feet, and Radius = 2,000 feet

Simulated groundwater source area	Number of particles from source area within model layer						Percentage of source area contributions within model layer					
	1	2	3	4	5	1–6	1	2	3	4	5	1–6
Northwest mountain-front boundary												
Big Lost River valley (BLR)	0	NA	NA	NA	NA	0	0.0	NA	NA	NA	NA	0.0
Little Lost River valley (LLR)	0	0	0	NA	NA	0	0.0	0.0	0.0	NA	NA	0.0
Birch Creek valley (BC)	0	0	0	0	NA	0	0.0	0.0	0.0	0.0	NA	0.0
Northwest mountain-front subtotal	0	0	0	0	NA	0	0.0	0.0	0.0	0.0	NA	0.0
Northeast regional-underflow boundary												
Reno Ranch section (Re)	0	0	0	0	0	0	0.0	0.0	0.0	0.0	0.0	0.0
Monteview section (Mo)	719	0	0	0	NA	719	94.9	0.0	0.0	0.0	NA	31.9
Mud Lake section (ML)	39	52	860	587	NA	1,538	5.1	100.0	100.0	100.0	NA	68.1
Terreton section (Te)	0	0	0	0	0	0	0.0	0.0	0.0	0.0	0.0	0.0
Northeast regional-underflow subtotal	758	52	860	587	0	2,257	100.0	100.0	100.0	100.0	0.0	100.0
Water table boundary												
Streamflow infiltration												
Big Lost River infiltration												
Stream reach 600–601	0	NA	NA	NA	NA	0	0.0	NA	NA	NA	NA	0.0
Stream reach 602–605	0	NA	NA	NA	NA	0	0.0	NA	NA	NA	NA	0.0
Stream reach 606–607	0	NA	NA	NA	NA	0	0.0	NA	NA	NA	NA	0.0
Stream reach 608–610	0	NA	NA	NA	NA	0	0.0	NA	NA	NA	NA	0.0
Little Lost River infiltration												
Stream reach 611	0	NA	NA	NA	NA	0	0.0	NA	NA	NA	NA	0.0
Birch Creek infiltration												
Stream reach 612	0	NA	NA	NA	NA	0	0.0	NA	NA	NA	NA	0.0
Streamflow-infiltration subtotal	0	NA	NA	NA	NA	0	0.0	NA	NA	NA	NA	0.0
Orphans	0	0	0	0	0	0	0.0	0.0	0.0	0.0	0.0	0.0
Total	758	52	860	587	0	2,257	100.0	100.0	100.0	100.0	0.0	100.0

Appendix G. Summary of backward particle-tracking results from a sensitivity analysis of the source area contributions to changes in input parameters for the layered-grid and NPR-W01 well simulations.—Continued

[Locations of source areas are shown in figure 3. Orphans are identified as particles that did not terminate in one of the specified source areas. None of the source areas include model layer 6. Streamflow reaches are located between the model boundary and streamflow-gaging station (map No.) 504 (600–601), within the Big Lost River spreading area (602–605), between streamflow-gaging stations 504 and 506 (606–607), within the Big Lost River sinks and playas (608–610), within the Little Lost River downstream of the model boundary (611), and within Birch Creek downstream of the model boundary (612). **FRAC:** the fraction of the total outflow from a cell contributed by an internal source; used to identify weak-source cells. **Layer:** the model layer particles are located in at their release. **Radius:** the aerial radius of the cylindrical volume defining particle release locations, in feet. **Depth interval:** the depths of released particles are distributed randomly in this depth interval, in feet below the top of model layer 1. **Abbreviation:** NA, not applicable]

NPR-W01 well simulation; FRAC = 0.5, Depth interval = 0–200 feet, and Radius = 5,000 feet

Simulated groundwater source area	Number of particles from source area within model layer						Percentage of source area contributions within model layer					
	1	2	3	4	5	1-6	1	2	3	4	5	1-6
Northwest mountain-front boundary												
Big Lost River valley (BLR)	0	NA	NA	NA	NA	0	0.0	NA	NA	NA	NA	0.0
Little Lost River valley (LLR)	269	0	0	NA	NA	269	3.8	0.0	0.0	NA	NA	3.0
Birch Creek valley (BC)	0	0	0	0	NA	0	0.0	0.0	0.0	0.0	NA	0.0
Northwest mountain-front subtotal	269	0	0	0	NA	269	3.8	0.0	0.0	0.0	NA	3.0
Northeast regional-underflow boundary												
Reno Ranch section (Re)	40	0	0	0	0	40	0.6	0.0	0.0	0.0	0.0	0.4
Monteview section (Mo)	1,200	0	0	0	NA	1,200	16.8	0.0	0.0	0.0	NA	13.3
Mud Lake section (ML)	333	94	663	1,120	NA	2,210	4.7	100.0	100.0	100.0	NA	24.5
Terreton section (Te)	0	0	0	0	0	0	0.0	0.0	0.0	0.0	0.0	0.0
Northeast regional-underflow subtotal	1,573	94	663	1,120	0	3,450	22.0	100.0	100.0	100.0	0.0	38.2
Water table boundary												
Streamflow infiltration												
Big Lost River infiltration												
Stream reach 600–601	0	NA	NA	NA	NA	0	0.0	NA	NA	NA	NA	0.0
Stream reach 602–605	0	NA	NA	NA	NA	0	0.0	NA	NA	NA	NA	0.0
Stream reach 606–607	1,683	NA	NA	NA	NA	1,683	23.5	NA	NA	NA	NA	18.6
Stream reach 608–610	2,455	NA	NA	NA	NA	2,455	34.3	NA	NA	NA	NA	27.2
Little Lost River infiltration												
Stream reach 611	0	NA	NA	NA	NA	0	0.0	NA	NA	NA	NA	0.0
Birch Creek infiltration												
Stream reach 512	187	NA	NA	NA	NA	187	2.6	NA	NA	NA	NA	2.1
Streamflow-infiltration subtotal	4,325	NA	NA	NA	NA	4,325	60.5	NA	NA	NA	NA	47.9
Orphans	984	0	0	0	0	984	13.8	0.0	0.0	0.0	0.0	10.9
Total	7,151	94	663	1,120	0	9,028	100.0	100.0	100.0	100.0	0.0	100.0

Appendix G. Summary of backward particle-tracking results from a sensitivity analysis of the source area contributions to changes in input parameters for the layered-grid and NPR-W01 well simulations.—Continued

[Locations of source areas are shown in figure 3. Orphans are identified as particles that did not terminate in one of the specified source areas. None of the source areas include model layer 6. Streamflow reaches are located between the model boundary and streamflow-gaging station (map No.) 504 (600–601), within the Big Lost River spreading area (602–605), between streamflow-gaging stations 504 and 506 (606–607), within the Big Lost River sinks and playas (608–610), within the Little Lost River downstream of the model boundary (611), and within Birch Creek downstream of the model boundary (612). **FRAC:** the fraction of the total outflow from a cell contributed by an internal source; used to identify weak-source cells. **Layer:** the model layer particles are located in at their release. **Depth interval:** the depths of released particles are distributed randomly in this depth interval, in feet below the top of model layer 1. **Radius:** the aerial radius of the cylindrical volume defining particle release locations, in feet. **Abbreviation:** NA, not applicable]

NPR-W01 well simulation; FRAC = 0.5, Depth interval = 0-200 feet, and Radius = 660 feet

Simulated groundwater source area	Number of particles from source area within model layer						Percentage of source area contributions within model layer					
	1	2	3	4	5	1–6	1	2	3	4	5	1–6
Northwest mountain-front boundary												
Big Lost River valley (BLR)	0	NA	NA	NA	NA	0	0.0	NA	NA	NA	NA	0.0
Little Lost River valley (LLR)	769	0	0	NA	NA	769	10.8	0.0	0.0	NA	NA	8.5
Birch Creek valley (BC)	0	0	0	0	NA	0	0.0	0.0	0.0	0.0	NA	0.0
Northwest mountain-front subtotal	769	0	0	0	NA	769	10.8	0.0	0.0	0.0	NA	8.5
Northeast regional-underflow boundary												
Reno Ranch section (Re)	16	0	0	0	0	16	0.2	0.0	0.0	0.0	0.0	0.2
Monteview section (Mo)	630	0	0	0	NA	630	8.8	0.0	0.0	0.0	NA	7.0
Mud Lake section (ML)	580	84	1,622	183	NA	2,469	8.1	100.0	100.0	100.0	NA	27.3
Terreton section (Te)	0	0	0	0	0	0	0.0	0.0	0.0	0.0	0.0	0.0
Northeast regional-underflow subtotal	1,226	84	1,622	183	0	3,115	17.2	100.0	100.0	100.0	0.0	34.5
Water table boundary												
Streamflow infiltration												
Big Lost River infiltration												
Stream reach 600–601	0	NA	NA	NA	NA	0	0.0	NA	NA	NA	NA	0.0
Stream reach 602–605	0	NA	NA	NA	NA	0	0.0	NA	NA	NA	NA	0.0
Stream reach 606–607	162	NA	NA	NA	NA	162	2.3	NA	NA	NA	NA	1.8
Stream reach 608–610	3,793	NA	NA	NA	NA	3,793	53.1	NA	NA	NA	NA	42.0
Little Lost River infiltration												
Stream reach 611	0	NA	NA	NA	NA	0	0.0	NA	NA	NA	NA	0.0
Birch Creek infiltration												
Stream reach 612	42	NA	NA	NA	NA	42	0.6	NA	NA	NA	NA	0.5
Streamflow-infiltration subtotal	3,997	NA	NA	NA	NA	3,997	56.0	NA	NA	NA	NA	44.3
Orphans	1,147	0	0	0	0	1,147	16.1	0.0	0.0	0.0	0.0	12.7
Total	7,139	84	1,622	183	0	9,028	100.0	100.0	100.0	100.0	0.0	100.0